Re-presenting Research

Florentine Marnel Sterk
Merel M. van Goch

Re-presenting Research

A Guide to Analyzing Popularization Strategies
in Science Journalism and Science Communication

Florentine Marnel Sterk
Institute for Language Sciences
Utrecht University
Utrecht, The Netherlands

Merel M. van Goch
Institute for Cultural Inquiry
Utrecht University
Utrecht, The Netherlands

ISBN 978-3-031-28173-0   ISBN 978-3-031-28174-7   (eBook)
https://doi.org/10.1007/978-3-031-28174-7

Cover pattern © Melisa Hasan

This Palgrave Macmillan imprint is published by the registered company Springer Nature Switzerland AG.
The registered company address is: Gewerbestrasse 11, 6330 Cham, Switzerland

ACKNOWLEDGMENTS

We want to thank the Open Access Fund of Utrecht University's Library, our research institutes, Institute for Language Sciences and Institute for Cultural Inquiry at Utrecht University, and our home base Liberal Arts and Sciences for contributing to the opportunity to publish this work open access.

We want to thank the members of the research groups Subjects in Interdisciplinary Learning & Teaching at the Institute for Cultural Inquiry (Utrecht University), and Language & Communication and Language & Education at the Institute for Language Sciences (Utrecht University), as well as members of the Science Communication Unit (University of West England Bristol) and the Interuniversity Centre for Educational Sciences for the fruitful discussions we have shared about popularization discourse.

We are particularly grateful to Cathy Scott for her amazing support and editorial assistance. We also thank two anonymous reviewers for their generous and enthusiastic feedback.

We thank Iris van der Tuin and Michael Burke for their valuable feedback.

Most of all, we thank the students who took part in our research. It is because of your newspaper articles that we were inspired to write this book.

CONTENTS

About the Authors

Florentine Marnel Sterk is Junior Assistant Professor at the interdisciplinary undergraduate program Liberal Arts and Sciences at Utrecht University, The Netherlands. She explores how popularization writing skills can best be taught in interdisciplinary university programs, to enable students to communicate effectively outside of their own academic niche and in interdisciplinary research settings. Furthermore, she aims to offer insights into effective popularization about research from interdisciplinary research settings. She also teaches courses on academic skills, writing skills, and interdisciplinary research methodology.

Merel M. van Goch is Assistant Professor at the interdisciplinary undergraduate program Liberal Arts and Sciences at Utrecht University, The Netherlands. Her work is driven by the motivation to provide people with the optimal circumstances to explore and foster their talents and interests. She is interested in how and what students and researchers learn, especially in interdisciplinary contexts, because interdisciplinarity magnifies the potential for learning. She studies metacognition, creativity, and other competences relevant to higher education, and her teaching is characterized by fostering students' self-directed learning.

LIST OF TABLES

CHAPTER 1

Introduction: The Re-Presentation of Research in Popularization Discourse

Abstract In this chapter, we introduce our book, *Re-presenting Research*. In this book, the focus is firmly on textual popularization discourse as a form of communication to re-present academic insights to a larger, non-expert audience in an understandable and engaging manner. Our own research into science journalism writing skills in undergraduate university students serves as a starting point for the exploration of popularization discourse as a genre. The book gives an exposition of current theories, employed methodologies, and existing frameworks and rubrics that cover popularization discourse. We focus on the conceptual and textual features of popularization discourse, which are called 'strategies' in this book. In this introduction, we establish the need for this book and introduce what it is about, the topics it will cover, and the example analyses that are provided.

Keywords Popularization discourse • Science communication • Science journalism • Textual features • Strategies • Re-presentation

Anyone who has ever tried to present a rather abstract scientific subject in a popular manner knows the great difficulties of such an attempt. Either he[1] succeeds in being intelligible by concealing the core of the problem and by

---

[1] In this book, we have used inclusive language. Note that in example texts or cited texts that are presented in this book, 'he' or 'she' might be used; in these cases we have not edited the original text.

F. M. Sterk, M. M. van Goch, *Re-presenting Research*,
https://doi.org/10.1007/978-3-031-28174-7_1

offering to the reader only superficial aspects or vague allusions, thus deceiving the reader by arousing in him the deceptive illusion of comprehension; or else he gives an expert account of the problem, but in such a fashion that the untrained reader is unable to follow the exposition and becomes discouraged from reading any further.

If these two categories are omitted from today's popular scientific literature, surprisingly little remains. But the little that is left is very valuable indeed. It is of great importance that the general public be given an opportunity to experience—consciously and intelligently—the efforts and results of scientific research. It is not sufficient that each result be taken up, elaborated, and applied by a few specialists in the field.
—Albert Einstein *(1948, as cited in Barnett, 1968, p. 9)*

Every day, new and exciting scientific findings are communicated to the general public, whether it be in the form of news reports, newspaper articles, YouTube videos, TikTok reels, or as part of science entertainment shows like MythBusters. Many people are eager to read up on new discoveries and developments in the academic world, such as the latest advances in the battle against COVID-19, the pictures of deep space shared by the James Webb Space Telescope, or the discovery of bacteria that are large enough to see with the naked eye. This growing corpus of texts and visuals enables us, as the general public, to have a good grasp of the elements that make scientific research interesting, and an idea about what makes certain stories more entertaining than others.

To give an example, in 2021, researchers from the Rosalind Franklin Institute and the University of Reading published their work on nanobodies, antibodies produced by the immune system of llamas that could be used in the treatment of SARS-CoV-2, in the journal *Nature Communications* (Huo et al., 2021). In newspaper articles that were written about this research, the main focus was not on nanobodies, but instead, Fifi the Llama was introduced. Fifi is a fun, fluffy, and interesting creature that appeals to our imagination and helps us understand the difficult subject matter of immunotherapy that is being discussed:

> By vaccinating Fifi with a tiny, non-infectious piece of the viral protein, the scientists stimulated her immune system to make the special molecules. The scientists then carefully picked out and purified the most potent nanobodies in a sample of Fifi's blood; those that matched the viral protein most closely, like the key that best fits a specific lock. (Gill, 2021)

Fifi the Llama enables the writer to draw the abstract information about the method that was used into the realm of the tangible. Fifi moves the story from the academic world of sterile laboratories into the everyday life of an actual animal. The addition of Fifi as a main character in the story of this research is what makes the newspaper article interesting to the reader. In other words, Fifi is the element that 'sells' the story.

For scientific findings and innovations to have an impact beyond the ivory tower—that is, to be noteworthy to society and influential on everyday life—they should be presented outside of the specific disciplinary community in which they were produced. To communicate in a way that is comprehensible and engaging to a large audience, findings from the academic world that are presented in academic papers or reports need to be distilled down to their core and presented in an attractive and understandable manner. In other words, *re-presentation of academic discourse* is needed, with the resulting product being called popularization discourse. When a researcher performs this re-presentation process for their own work, the outcome is called *science communication*. More commonly, though, a journalist re-presents academic work and by doing so constructs *science journalism*. Both science communication and science journalism fall under the umbrella of popularization discourse. (See the Glossary at the end of the book for definitions of important terms.)

To achieve the goals of popularization discourse, writers can use textual features; the use of an everyday life example in the form of Fifi the Llama is one of them, but there are many more. Through these textual features, or 'strategies,' the distinct *genre* of popularization discourse is constructed.

Popularization discourse is an important text genre both for the academic world and for the general audience. But why should we improve our understanding of science communication and science journalism? Science communication and science journalism fulfill an indispensable role in translating academic discourse into the realm of society and everyday life, and in bridging the gap between scientific advances at large versus the individual person who is influenced by them. Many scientific topics lead to societal debate or even controversy, which then reflects on how we as a society think and feel about academic research. Controversy surrounding COVID-19 vaccines, artificial intelligence, and climate change are some of the topics that have recently sparked heated debates. As Myers put it: "We cannot understand why there are tensions about genetically modified organisms, vaccinations, or climate change if we assume that science is distinct from the rest of culture, and that the public is, on scientific matters, a blank slate" (2003, p. 274). Although the research fields of science communication and science journalism have existed for several decades, controversy surrounding research findings is one of the reasons why it remains important to discuss the discourse from an academic stance.

1.1   WHY THIS BOOK?

Many academic studies today are devoted to the theoretical discussion of the nature of popularization, and to research versus society. Yet, only very few studies are devoted to popularization discourse *as a text genre*. Analysis of popularization discourse could be the first step in assessing and improving it. Still, a theoretically grounded and empirically tested framework to analyze popularization discourse is missing from the literature. A pressing issue is that research about the textual features that make up the genre of popularization discourse is scarce. Although some exceptions to this rule do exist, overall there is a structural lack of development of analytical frameworks that can be used to systematically *analyze the use of textual features*. This is hampering our academic knowledge about popularization discourse, and consequently, our real-world and lived knowledge too.

This issue became evident when we started our research into the science journalism writing strategies of undergraduate students. We strived to analyze their strategies in the most reliable, objective, and research-informed manner as possible, but we quickly realized that the tool we needed—an analytical framework suitable for our purposes—did not yet exist. In the academic literature, we were able to find a handful of frameworks that describe the genre of popularization discourse, but they were all static *results* of ad hoc analyses of one specific text form or topic within the genre. In other words, none of these frameworks were meant *to be employed in the analysis of other texts*. We thus set out to make our own analytical framework that would be usable in a wider context than merely our own specific research interests. The development of our framework was a true learning process; we encountered many hurdles and unknowns. We both have a background in language-related disciplines and have ample experience in mono- and interdisciplinary education, coupled with a keen research interest in research methodology and educational assessment. During the development of the framework, we had stimulating and insightful discussions, not just about the content of the framework itself and its process of development, but also on a more overarching level about text analysis and communication sciences on the one hand, and education, educational sciences, pedagogy, and didactics on the other. "Someone should write about this!" we exclaimed more than once. And thus, the idea of this book was born.

The little insight that exists in the academic literature comes from frameworks or rubrics describing the textual features of popularization discourse, or popularization strategies. Frameworks—analytic codebooks that display textual features and their properties—are developed through text analysis, yet the insights they produce are often contextualized within a subgenre—such as the opinion editorial—and a specific field of study, usually the natural sciences. Rubrics—analytic tables that show assessment criteria and grading points and can be used to assess student performance—are mostly based on literature reviews instead of empirical insights and developed with the aim of assessing popularization discourse in educational settings only. Rubrics offer a text-as-a-whole view, and frameworks a zoomed-in view, on the level of textual strategies. Components from these rubrics and frameworks can provide insight into popularization strategies. However, there is no consensus between sources. On an individual level, none of these existing frameworks or rubrics can give a complete or overarching overview, or are developed with the aim of using them as coding schemes in future studies outside of the specific context for which they were developed or with multiple coders in mind. An analytical framework that overarches academic disciplines, all subgenres of written popularization discourse, and all potential popularization strategies was still missing until now. Ideally a framework should be adaptable to various uses, contexts, and situations, for example usable in the analysis of a corpus of newspaper articles as well as being a learning tool in educational settings across disciplines – and to be able to produce robust results when used by multiple raters.

As far as we know, this book is unique in its combination of the disciplinary perspectives of language and education, and of theory, methods, and applicability. In terms of theory, this is one of the first works to focus on popularization discourse *as a discourse type*. Existing works focus on science communication as a communicative type or a communicative activity, thereby focusing purely on effectiveness and overlooking the discourse aspect. Regarding methodology, *Re-presenting Research* is unique because while the literature does contain some studies that offer insight into discourse strategies used in popularization, none of these works presents an *analytical* framework as a tool to analyze popularization discourse.

Our framework is empirically grounded: it is based on the literature and direct observations. With respect to application, this book offers three examples of the applicability of our framework and discusses considerations for readers who want to develop their own framework.

1.2   What the Book Is About

The central focus of this book is the re-presentation of scientific findings in popularization discourse. This book is called *Re-presenting Research* because popularization discourse fulfills two different roles. One the one hand it represents research in the sense that popularization discourse is used as a 'spokesperson' for research in communication to a broad audience. On the other hand, popularization discourse re-presents research; it finds a new context through which to frame the academic findings and does so through remodeled language that is suitable for a broad audience. We specifically focus on the textual features, or strategies, which make up popularization discourse. The re-presentation of academic discourse into popularization discourse involves two processes: recontextualization and reformulation. *Recontextualization* revolves around reimagining findings within an everyday context. Changes take place on a conceptual level and the focus is on newsworthiness and applicability of scientific findings. *Reformulation*, on the other hand, has the main aim of heightening engagement with the text and increasing text comprehensibility. Reformulation causes changes on a linguistic and textual level. Together, recontextualization and reformulation remodel academic discourse into newsworthy and understandable popularization discourse.

Popularization discourse broadly consists of science communication and science journalism. Science communication and science journalism are two fields that are in large part centered around the natural sciences and STEM fields (science, technology, engineering, and math). The terms themselves already provide a clue to this hyperfocus: 'science' communication and 'science' journalism. It is only more recently that other fields such as the social sciences and humanities (the SSH field) have also been included in popularization efforts. In fact, communication to a broad audience from other fields than the natural sciences has its own term: research communication. Social sciences are often overlooked in these efforts to characterize the different forms of communication about academic results. This might be because they are assumed to be part of the 'science' in science communication and science journalism, while in practice social sciences are often not recognized as such from the perspective of communication efforts from the STEM fields (Wilkinson & Weitkamp, 2016). Furthermore, in multidisciplinary and interdisciplinary research, insights from different disciplinary fields are integrated to form one comprehensive view of the research problem. For those studies, popularization must include the combination of multiple strategies to effectively communicate about the different disciplinary insights as well as the integrated conclusion (see Sterk, 2023). In *Re-presenting Research*, we aim to give insight into the textual features of popularization of the entire academic field, not just of the natural sciences. The analytical framework that we will present in this book, therefore, is constructed with the aim of being applicable to all disciplinary as well as multidisciplinary and interdisciplinary specializations.[2]

Re-presenting Research is mainly written for researchers and/or educators looking into the analysis of popularization discourse. The book focuses on *the doing* and *the being* of popularization discourse and acts as a guide for those working with or on popularization discourse–whether it is to analyze, write, or learn about it.

1.3   AIMS AND SCOPE OF THE BOOK

In this book, we focus on the structural and componential analysis of written popularization texts or, in other words, the use of *strategies*. In doing so we follow the definition by August et al. (2020), who refer to writing strategies as theory-driven communicative goals that consist of lexical to multi-sentence features. *Re-presenting Research* presents a review of current theories about popularization discourse and existing frameworks describing it. We then elaborate on the construction and validation of a

---

[2] In this book, we use the term 'research' to denote the entire field of academic discovery, which includes the humanities, social sciences, and natural sciences. We will use 'sciences' or 'natural sciences' to refer to the STEM field specifically. Likewise, we use 'researcher' as an all-encompassing term for those academics who perform research in all academic fields. Note that 'scientist' is used in some of the cited texts that are presented in this book; in these cases we have not edited the original text. The discourse of the academic world is referred to as 'academic discourse,' but still appears as 'scientific discourse' in cited sources. Even though science communication and science journalism often refer specifically to communication about the natural sciences, in this book we use them as all-encompassing terms to mean communication about all the disciplinary fields as well as multidisciplinary and interdisciplinary fields.

new empirically grounded analytical framework, based on the literature and direct observations, that is:

1. usable in any subgenre of popularization discourse
2. usable in disciplinary but also multidisciplinary and interdisciplinary settings
3. reliable for use with multiple raters
4. easy to apply by offering application remarks and explanations of strategies

We present three examples of how the analytical framework can be used to analyze texts written by authors with various levels of experience. We have included texts from both professional and student writers, to show that the framework can supply valuable insights in the different contexts of both academic and educational settings. *Re-presenting Research* therefore adds to the methodology of the fields of science communication, discourse analysis, and communication studies.

The book is focused on written popularization discourse and thus does not cover spoken, visual, or interactive science communication or science journalism, because our aim was to produce a brief text acting both as a guide for analyzing science communication and science journalism texts and as an updated review of the literature in a field that is still underexplored from a linguistic point of view. Our guiding principle was for this book to be practical and applicable; therefore, the theoretical chapters have been kept brief and to the point, and readers are referred to additional literature where applicable.

Re-presenting Research is for anyone interested in writing about research for the general public. The main audience we had in mind while writing is that of academics, who can gain knowledge about popularization discourse and find out how to analyze it. Furthermore, students, science communicators, science journalists, and practitioners can use this book. Undergraduate and graduate students can hone their science communication skills by learning how to write in an appealing way to a broad audience about their own research projects. Science communicators and practitioners at universities or academic associations, companies, or governments will learn more about the discourse type, helping them to communicate about scientific findings constructed by others. In education, the book can be used to offer science communication components in undergraduate or graduate thesis writing courses, in any academic field. *Re-presenting Research* can also be used in graduate programs in science communication or science journalism that train students to become science communicators or science journalists and that typically offer specific writing courses.

1.4 OUTLINE OF CHAPTERS

Re-presenting Research is divided into two sections. The first section provides the theoretical and methodological background. The second section consists of three examples of using our framework as an analytical tool. Chapter 2 offers theoretical grounding of the framework for popularization discourse. It presents theoretical background knowledge into popularization discourse, science communication, and science journalism as well as insight on recontextualization and reformulation as textual construction processes, and on challenges in research and practice. Chapter 3 discusses methodological considerations in analyzing popularization discourse. The chapter presents an overview of existing frameworks and rubrics to analyze or assess popularization texts. Chapter 4 introduces our analytical framework for popularization discourse. The development of the framework is discussed. The framework, its five themes (Subject Matter, Tailoring Information to the Reader, Credibility, Stance, and Engagement), and its 34 strategies are reviewed, and we explain how to code the strategies when using the framework. We also refer the reader to additional literature on each strategy. Chapter 5 provides the first example of using the framework as an analytical tool. The chapter shows how the framework can be used to thoroughly analyze one science journalism text and which insights can be gained from the analysis. Chapter 6 presents the second example of how the framework can be used. The framework is applied to a corpus of professional science journalism texts published in popular media. The analysis focused on which of the strategies were used by professional science journalists, with the goal of providing insights into the genre of science journalism. Chapter 7 is the third example of using the framework. In this chapter, the framework is used to analyze a corpus of newspaper articles written by first-year undergraduate liberal education students, to answer questions about *how often* (quantitative) and *how* (qualitative) the strategies in the framework are used by these student writers. Chapter 8 brings together the theoretical insights discussed in

Chaps. 2 and 3, and the practical and applied insights gained from the use of the framework as an analytic tool in Chaps. 5 through 7. It also discusses the ample opportunities for future research that follow from the contents of this book, and ends with advice for readers wanting to construct their own framework.

Re-presenting Research is best read cover to cover. Readers interested in the theoretical and methodological background of popularization discourse will find that Chaps. 2 and 3 can also be read stand-alone. Readers interested in developing their own analytical framework can start at Chap. 4 and review Chaps. 5 through 7 for applied examples. Readers who are interested in using the analytical framework to analyze or produce popularization texts are best advised to read through all chapters. We wrote the book we would have liked to have read when we started our research into science journalism writing skills in first-year students. Our vision for this work is to encourage the reader to either use our framework to analyze popularized texts or develop their own unique framework.

References

August, T., Kim, L., Reinecke, K., & Smith, N. A. (2020). Writing strategies for science communication: Data and computational analysis. *Proceedings of the 2020 Conference on Empirical Methods in Natural Language Processing*, 5327–5344. https://doi.org/10.18653/v1/2020.emnlp-main.429

Barnett, L. (1978). *The universe and dr. Einstein* (Rev. ed.). Bantam Books.

Gill, V. (2021, September 22). Covid: Immune therapy from llamas shows promise. *BBC*. https://www.bbc.com/news/science-environment-58628689

Huo, J., Mikolajek, H., Le Bas, A., Clark, J. J., Sharma, P., Kipar, A., & Owens, R. J. (2021). A potent SARS-CoV-2 neutralising nanobody shows therapeutic efficacy in the Syrian golden hamster model of COVID-19. *Nature Communications, 12*(5469). https://doi.org/10.1038/s41467-021-25480-z

Myers, G. (2003). Discourse studies of scientific popularization: Questioning the boundaries. *Discourse Studies, 5*(2), 265–279. https://doi.org/10.1177/1461445603005002006

Sterk, F. M. (2023). Wetenschapscommunicatie van interdisciplinaire onderzoeksprojecten [Popularization of interdisciplinary research projects]. *Tijdschrift voor Hoger Onderwijs*.

Wilkinson, C., & Weitkamp, E. (2016). *Creative research communication: Theory and practice*. Manchester University Press.

CHAPTER 2

Theoretical Considerations: Recontextualization and Reformulation in Popularization Discourse

Abstract This chapter establishes popularization discourse as a genre. In doing so, a distinction can be made between the act of popularization itself and the resulting product of popularization discourse in the form of texts. Popularization draws academic findings into the realm of society and everyday life, which means the focus shifts from the researched phenomenon itself to claims about newsworthiness and the application of findings. The chapter also discusses the two main subgenres of science communication and science journalism, which differ mostly in terms of who produces the discourse, that is, a researcher versus a journalist. The positionality of academic discourse versus popularization discourse is discussed, as the two discourses used to be seen as distinct genres but should be considered as different points on a continuum. The focus is on reformulation and recontextualization as the two main processes in the textual construction of popularization discourse.

Keywords Popularization discourse • Science communication • Science journalism • Reformulation • Recontextualization

2.1 Introduction

In this first chapter, background information is provided to offer insight into the context in which our framework for popularization discourse is situated. It discusses popularization discourse, science communication

© The Author(s) 2023
F. M. Sterk, M. M. van Goch, *Re-presenting Research*,
https://doi.org/10.1007/978-3-031-28174-7_2

and science journalism, older and current models, recontextualization and reformulation as textual construction processes, and the challenges that are faced within the research field and practical field.

2.2   Popularization Discourse

Popularization is a concept with two meanings; it is both a verb and a noun. On the one hand, it can refer to the act of giving insight into academic findings in an understandable and engaging way for a non-academic or non-expert audience, that is, the act of transformation (Calsamiglia & Van Dijk, 2004). On the other hand, it can refer to the discourse that is produced because of that act, which can include multiple discursive-semiotic practices and a multitude of media forms (Calsamiglia & Van Dijk, 2004). To avoid confusion, in this book we use the verb 'popularization' to mean the act, and the noun 'popularization discourse' to denote the product.

In the academic literature, popularization is discussed from multiple research perspectives: applied linguistics, rhetoric, communication sciences, media studies, history, and science (Myers, 2003). In the field of discourse analysis, multiple views of popularization can be distinguished: as a translation or reformulation of academic discourse into a second discourse, as a discursive genre, as recontextualization, as dependent on processes in the media, or as a form of dialogic relationship between the scientific context and other contexts (Grillo et al., 2016). Popularization discourse is characterized primarily through the context and social situation in which it is constructed. What matters most are the actors, their role in the communication, their knowledge, purposes, and beliefs, and the applicability of their knowledge in everyday life (Calsamiglia & Van Dijk, 2004). According to Calsamiglia (2003), "these days the scientific disciplines express themselves in what for the non-specialist is an unknown, hermetic and difficult language" (p. 141). Through popularization, knowledge produced in these academic and specialized practices is transformed into knowledge for a lay audience: the resulting discourse connects to everyday life and is written in understandable language (Calsamiglia & Van Dijk, 2004; Gotti, 2014). Or, as Hyland framed it:

> This [popular science] is a discourse related to the academy, its work, and its forms of communication but stripped of its more forbidding rhetorical features. While attempting to wield the authority of science, both scientific

facts and the argument forms of professional science are transformed in the process. (Hyland, 2010, p. 118)

In this quote, Hyland placed the focus on the adaption of rhetorical features in the process of popularization. At the same time, tension exists between the discourse of research and discourse of public communication, as Baram-Tsabari and Lewenstein pointed out: "Unfortunately, the two discourses are sometimes in tension: One rewards jargon, the other penalizes it; one rewards precision, the other accepts approximation; one rewards quantification, the other rewards storytelling and anecdotes" (Baram-Tsabari & Lewenstein, 2013, p. 80).

Even though popularization discourse is based on academic discourse, it is often defined by its differences to or adaptations from it. Popularization discourse brings the worlds of research and of society and everyday life closer together. Yet the different regard academic discourse and popularization discourse possess for scientific objects forms an obstacle in the 'translation' of academic findings from one discourse to the other:

> [F]or the former (the scientists) the object has an immanent value in scientific and specialist contexts. For the latter (the public) the value is external to all the theories and methods: what is important is its application, its utility, and the consequences of its use in people's lives. (Calsamiglia, 2003, p. 140)

This difference in positionality leads to one of the main themes in the process of constructing popularization discourse: the genre shift from scientific reports, which establish the validity of the observations, to science journalism texts, which celebrate academic research and underpin its significance. This is what Myers described in the following way:

> The professional articles create what I call a narrative of science; they follow the argument of the scientist, arrange time into a parallel series of simultaneous events all supporting their claim, and emphasize in their syntax and vocabulary the conceptual structure of the discipline. The popularizing articles, on the other hand, present a sequential narrative of nature in which the plant or animal, not the scientific activity, is the subject, the narrative is chronological, and the syntax and vocabulary emphasize the externality of nature to scientific practices. (1990, p. 142 as quoted in Giannoni, 2008)

Myers positioned academic texts with a focus on the researcher and research, whereas popularized texts focus on the phenomenon itself.

Fahnestock talked about the same phenomenon and described the two main rhetorical arguments in science journalism to be those of 'the wonder' and 'the application.' Concurrently, the rhetorical life of scientific observations cycles from the nature of a phenomenon toward its values and consequences in everyday life (Fahnestock, 1986). Scientific findings travel along a 'communicative path' that moves from the intraspecialistic (disciplinary) stage to the interspecialistic (academic, non-disciplinary) stage to the pedagogical and finally the popular stage. In each of these stages, details and meaning are removed and findings are solidified as (simple) facts (Bucchi, 2008). The presented information therefore changes: uncertainty is removed and direct quotes from authors are added (Fahnestock, 1986). In this process, scientific research that is conducted to gain more understanding of a phenomenon ('the wonder') is transformed and re-presented in a way that connects with society and everyday life ('the application'), while the focus shifts from the producer of the knowledge (the researcher) in the academic discourse back to the phenomenon as it appears in nature in the popularization discourse. These changes are made in order to make the presented information interesting and usable for a broad audience.

2.3   SCIENCE COMMUNICATION VERSUS SCIENCE JOURNALISM

Popularization discourse consists of two main subgenres: science communication and science journalism. In science communication, researchers communicate about findings from their own research. Although in some settings this is assumed to be one-way communication, other definitions assume a broader spectrum of communicative activities (Bultitude, 2011). An example is the definition from Burns et al., who viewed science communication as a culmination of four types of expertise: the use of appropriate skills, media, activities, and dialogue. Science communication may also be practiced by practitioners, mediators, and members of the general public. The aim of science communication is to elicit a personal response to research in the form of awareness, enjoyment, interest, opinion, or understanding (Burns et al., 2003, p. 191). Effective science communication is also defined as a means to inform the public to facilitate decision making (Fischhoff, 2013). Science communication includes, but is not the equivalent of, public awareness of science, public understanding of science, scientific literacy, and scientific culture (Burns et al., 2003). Science communication can be found on an institutional level, where motivations mostly lie on a utilitarian, economic, cultural, and democratic level, or on the individual level of the researcher, who can use science communication to boost their career or network, to develop new skills, or to obtain additional funding (Bultitude, 2011). Apart from written science communication, there are also plenty of spoken and interactive science communication options, such as lectures, science cafes, and festivals—although we will not specifically address these in this book.

In science journalism, a journalist—who has usually received extensive academic training—communicates about academic insights published by researchers in a clear and appealing way (Molek-Kozakowska, 2015). In the nineteenth century, popularization used to be part of a researcher's job, but this changed in the twentieth century when ongoing demarcation of science away from society meant that pursuing popularization activities could destroy a researcher's career. Mass media, on the other hand, had an unwavering interest in stories about academic research (although they were not always considered as such, specifically), which meant that journalists took on the popularization function. Even nowadays, when popularization is once again a popular activity for researchers, science journalists are still responsible for a large number of the stories about academic advancements (Dunwoody, 2014).

Science journalism is a mixed discourse of informative and explanatory elements. It explains research findings but also places them in the framework of public concern (Gotti, 2014). Its *interdiscursivity* is formed through three types of discourse: academic discourse, journalistic discourse, and pedagogical discourse. The latter is used as a learning tool, to provide scaffolding for readers to grasp new scientific information (Motta-Roth & Scherer, 2016). Science journalism, therefore, means not only a recontextualization of the research presented but also a framing and interpretation through which the public understanding is influenced (Molek-Kozakowska, 2015). Furthermore, features from both journalism and science communication are used. This combining of discourses poses some difficulties, as they do not share the exact same goal: where science communication primarily tries to offer a credible presentation of the research, journalism includes a double appeal of both remarkable science and appealing news stories (Molek-Kozakowska, 2015).

Another complicating factor is that although professional journalists are responsible for the re-presentation of research, they, in turn, are part of and dependent on the institutional organization of the media (Gotti, 2014). Newsworthy stories are more likely to be profitable, which means stories that score well on timeliness, impact, and proximity are more likely to be picked up (Molek-Kozakowska, 2015, 2017). It is exactly this pursuit of newsworthiness that might put the focus too much on infotainment and hamper the understanding of the actual research (Molek-Kozakowska, 2017). This also explains why science journalism faces criticism of "inaccuracy, sensationalism, oversimplifications and failing to engage audiences in meaningful debate about scientific issues"—although these claims are also contested (Secko et al., 2013, p. 62).

2.4   VIEWS AND MODELS OF POPULARIZATION DISCOURSE

Across time, different views and models have existed about popularization discourse. One of the earliest views on popularization was the culturally dominant or canonical view, in which popularization discourse only included texts for non-specialists (Myers, 2003). Popularizations were seen as a simplified and often degraded version of scientific knowledge, and were presented in stark contrast with "pure, genuine scientific knowledge" (Hilgartner, 1990, p. 519). Myers explained the canonical view in the following way: "Popularization includes only texts about science that are not addressed to other specialist scientists, with the assumption that the texts that are addressed to other specialists are something else, something much better: scientific discourse" (Myers, 2003, p. 265).

This way of thinking of popularization discourse as a less worthy, toned down version of academic discourse creates a divide between the two discourses. Consequently, it also leads to a gap between the academic and non-academic world. Indeed, the canonical view was politicized to demarcate research as only accessible for academic experts and to give those experts authority (Hilgartner, 1990). In doing so, a chasm was created between experts and lay people. Connected to the culturally dominant view of popularization are dissemination models, the most well known of which is the deficit model, which has been heavily critiqued since. The deficit model assumes that lay people (or 'the public') show skepticism or resistance to academic research because of a lack of knowledge. Information is the solution to this problem, which researchers can provide in a unilateral way. Popularization is very much seen as a pedagogical function in this model (Besley & Tanner, 2011; Miller, 2001; Trench, 2008).

Newer views do not position popularization discourse and academic discourse as distinct discourses (Myers, 2003). Academic discourse and popularization discourse are rather seen as genres on a continuum, with many different variances in between: "Only from the outside, and from a great distance, does scientific discourse seem to employ a single unified register" (Myers, 2003, p. 270). Or, to put it differently: "Popularization is a matter of degree" (Hilgartner, 1990, p. 528). This means that there is no one set form or mode of popularization discourse, and that there is no clear demarcation of where academic discourse ends and popularization discourse begins. Newer views see popularization discourse as more broadly applicable than just as a way of conveying information; it also gives center stage to persons, identities, experiences, and interaction. Therefore, popularization discourse also raises questions about the actors, institutions, and forms of authority involved (Myers, 2003). Hyland introduced the idea of *proximity* to explain "… the ways writers manage their display of expertise and interactions with readers through rhetorical choices that contextually construct both the writer and reader as people with similar understandings and goals" (Hyland, 2010, p. 116).

In doing so, the divide between the writer as 'expert' and reader as 'layperson' is fading, with both parties forming an integral part in the communication. The focus of popularization discourse, then, moves away from the pedagogical function and onto explaining the social stakes of issues involved (Moirand, 2003). Models connected to this view are those of dialogue and conversation. In dialogue models, different target audiences with their own background, information, and needs are taken into consideration. Communication based on these models is often still a one-way process, but the aim is to create two-way communication between researchers and their audience, and interaction plays a bigger role (Trench, 2008). Conversation models operate on the idea that researchers and the audience can work together to shape the communication, and non-academic voices and information are taken into consideration. In doing so, a three-way form of communication is created (Besley & Tanner, 2011; Miller, 2001; Trench, 2008).

In the early 2000s, ideas about popularization discourse became more centered on social representation of scientific knowledge, which in turn became largely mediated by the news media (Calsamiglia, 2003). Journalists recreate the original discourse within a new situation, which means actors and institutions involved in research get a different degree of authoritativeness assigned to them. Consequently, mass media become an active creator in the production of knowledge, insights, and opinions about research (Gotti, 2014). The media also show the influence of research on everyday life, its social or human side, and ways in which it can be (ab)used within society (Calsamiglia & Van Dijk, 2004).

Since the early 2000s, the landscape in which science communicators and science journalists operate has changed dramatically. With the move toward online media, audiences read and write about academic advances on (science) blogs and online media outlets. Researchers can directly communicate with audiences through Twitter (Brossard & Scheufele, 2013). Search engine algorithms create a latent bias in results, determining which information a user is able to find (Van Dijck, 2010). Online social networks further determine the information users are presented (Brossard, 2013; Brossard & Scheufele, 2013). The mainstream media used gatekeeping to judge what to present as news, yet online citizen journalists now use gatewatching to identify which news stories can be republished or reinterpreted, adding perspective to the story, and expanding coverage (Bruns, 2018). These changes no doubt call into question the changing power dynamics between researchers, the media, and the public. Bruns also noted that more recently, ideas have started appearing around filter bubbles and echo chambers "... that are each subject to their own internal 'groupthink', and no longer find the common communicative ground to sustain broader public debate and deliberation" (Bruns, 2018, p. 325). This shows that the media landscape in which science communicators and science journalists now operate is vastly different from 20 or even 10 years ago.

2.5   REFORMULATION AND RECONTEXTUALIZATION

As described above, the main characteristics that ultimately form popularization discourse are the social properties of its communicative context (Calsamiglia & Van Dijk, 2004), and the focus on application and consequences (Fahnestock, 1986). Two processes that are responsible for the *textual construction* of popularization discourse are reformulation and recontextualization (Bondi et al., 2013; Calsamiglia & Van Dijk, 2004; Ciapuscio, 2003; Gotti, 2014).

In recontextualization, part of the discourse is taken from one communicative context and re-presented in another one. For scientific facts, this means a move from an expert context to a lay context (Bondi et al., 2013). Recontextualization, then, entails presenting specialized knowledge in such a way that non-specialized readers can construct and integrate it into their frame of reference (Calsamiglia & Van Dijk, 2004). It might occur on an intratextual, intertextual, or interdiscursive level and can include changes in meaning or content (Bondi et al., 2013). The recontextualization of knowledge works on the level of changes in the cognitive dimension (established versus new knowledge), the situational dimension (interests, intentions, and purposes of writer and reader), and the social dimension (that is, the research process translated into a journalistic genre) (Calsamiglia, 2003). As Hall et al. defined it: "... recontextualization amounts to putting something in a different context and, by doing so, creating a new context for it" (1999 as cited in Ciapuscio, 2003, p. 210).

Recontextualization thus enables non-specialized readers to construct a non-specialist version of the specialized knowledge and integrate it into their frame of reference. For a writer, recontextualization helps to adapt to the constraints of the communicative events in which the popularization discourse appears (Calsamiglia & Van Dijk, 2004). Recontextualization strategies include forms of explanation such as definition, examples, and metaphors to link new knowledge to the reader's existing knowledge (Calsamiglia & Van Dijk, 2004), narratives, imagery, and specific expressive functions such as example, definition, denomination, description, exemplification, generalization, paraphrase or reformulation (Gotti, 2014), simplification or condensation, refocusing, expansion, and elaboration (Bondi et al., 2013). Differences between academic texts and popularized texts can be found in textual form, sentence subjects, grammatical voice, verb choices, modality, hedging, and rhetorical structure (Gotti, 2014). Through recontextualization, researchers are presented as actors in a "discovery event," that is, in a direct meeting between researcher and nature (Hyland, 2010, p. 126).

On the other hand, reformulation is a process that does not alter the content or context of the message but does remodel the language to a new target audience by using strategies like metaphor, simile, and figurative language. It is a process that is similar to intralinguistic translation, that is, translation within the same language. Reformulation is often assumed to be the only process in popularization, meaning the language of academic discourse would be adapted but the content would not. This is an assumption that fails to take recontextualization into consideration (Gotti, 2014).

What You Have Learned in This Chapter

- Popularization discourse used to be seen as separate from and less worthy than academic discourse, but in fact both are part of the same discourse.
- Popularization is a re-presentation process in which the focus shifts from 'the wonder' about nature to 'the application' of findings.
- Science communication is constructed by researchers, while science journalism is a hybrid discourse constructed by (knowledgeable) science journalists.
- Views and models about popularization have shifted over time, away from deficit models toward participation models.
- Recontextualization (creating a new context for the information) and reformulation (remodeling on a textual level) are the two main processes through which popularization discourse is constructed.

REFERENCES

Baram-Tsabari, A., & Lewenstein, B. V. (2013). An instrument for assessing scientists' written skills in public communication of science. *Science Communication, 35*(1), 56–85. https://doi.org/10.1177/1075547012440634

Besley, J. C., & Tanner, A. H. (2011). What science communication scholars think about training scientists to communicate. *Science Communication, 33*(2), 239–263. https://doi.org/10.1177/1075547010386972

Bondi, M., Cacchiani, S., & Mazzi, D. (2013). *Discourse in and through the media: Recontextualizing and reconceptualizing expert discourse.* In M. Bondi, S. Cacchiani, & D. Mazzi (Eds.), *Discourse in and through the media: Recontextualizing and reconceptualizing expert discourse* (pp. 1–21). Cambridge Scholars Publishing.

Brossard, D. (2013). New media landscapes and the science information consumer. *PNAS, 110*(3), 14096–14101. https://doi.org/10.1073/pnas.1212744110

Brossard, D., & Scheufele, D. A. (2013). Science, new media, and the public. *Science, 339*(6115), 40–41. https://doi.org/10.1126/science.1232329

Bruns, A. (2018). *Gatewatching and news curation: Journalism, social media, and the public sphere. Peter Lang.* https://doi.org/10.3726/b13293

Bucchi, M. (2008). Of deficits, deviations and dialogues: Theories of public communication in science. In M. Bucchi & B. Trench (Eds.), *Handbook of public communication of science and technology* (pp. 57–76). Routledge.

Bultitude, K. (2011). The why and how of science communication. In P. Rosulek (Ed.), *Science communication*. Pilsen.

Burns, T. W., O'Connor, D. J., & Stocklmayer, S. M. (2003). Science communication: A contemporary definition. *Public Understanding of Science, 12*, 183–202. https://doi.org/10.1177/09636625030122004

Calsamiglia, H. (2003). Popularization discourse. *Discourse Studies, 5*(2), 139–146. https://doi.org/10.1177/1461445603005002307

Calsamiglia, H., & Van Dijk, T. A. (2004). Popularization discourse and knowledge about the genome. *Discourse & Society, 15*(4), 369–389. https://doi.org/10.1177/0957926504043705

Ciapuscio, G. E. (2003). Formulation and reformulation procedures in verbal interactions between experts and (semi-)laypersons. *Discourse Studies, 5*(2), 207–233. https://doi.org/10.1177/1461445603005002004

Dunwoody, S. (2014). Science journalism: Prospects in the digital age. In M. Bucchi & B. Trench (Eds.), *Routledge handbook of public communication of science and technology* (2nd ed., pp. 27–39). Routledge. https://doi.org/10.4324/9780203483794.ch3

Fahnestock, J. (1986). Accommodating science: The rhetorical life of scientific facts. *Written Communication, 3*(3), 275–296. https://doi.org/10.1177/2F0741088386003003001

Fischhoff, B. (2013). The sciences of science communication. *Proceedings of the National Academy of Sciences, 110*(suppl. 3), 14033–13039. https://doi.org/10.1073/pnas.1213273110

Giannoni, D. S. (2008). Popularizing features in English journal editorials. *English for Specific Purposes, 27*(2), 212–232. https://doi.org/10.1016/2006.12.001

Gotti, M. (2014). Reformulation and recontextualization in popularization discourse. *Ibérica, 27*, 15–34.

Grillo, S. V. C., Giering, M. E., & Motta-Roth, D. (2016). Discourse perspectives of science divulgation/Popularization perspectivas discursivas da divulgação/popularização da ciência. *Bakhtiniana, 11*(2), 3–13. https://doi.org/10.1590/2176-457327166

Hilgartner, S. (1990). The dominant view of popularization: Conceptual problems, political uses. *Social Studies of Science, 20*(3), 519–539. https://doi.org/10.1177/030631290020003006

Hyland, K. (2010). Constructing proximity: Relating to readers in popular and professional science. *Journal of English for Academic Purposes, 9*(2), 116–127. https://doi.org/10.1016/j.jeap.2010.02.00

Miller, S. (2001). Public understanding of science at the crossroads. *Public Understanding of Science, 10*(1), 115–120. https://doi.org/10.1088/0963-6625/10/1/308

Moirand, S. (2003). Communicative and cognitive dimensions of discourse on science in the French mass media. *Discourse Studies, 5*(2), 175–206. https://doi.org/10.1177/1461445603005002003

Molek-Kozakowska, K. (2015). Pragmalinguistic categories in discourse analysis of science journalism. *Lodz Papers in Pragmatics, 11*(2), 157–179. https://doi.org/10.1515/lpp-2015-0009

Molek-Kozakowska, K. (2017). Communicating environmental science beyond academia: Stylistic patterns of newsworthiness in popular science journalism. *Discourse & Communication, 11*(1), 69–88. https://doi.org/10.1177/1750481316683294

Motta-Roth, D., & Scherer, A. S. (2016). Popularização da ciência: A interdiscursividade entre ciência, pedagogia e jornalismo [Science popularization: Interdiscursivity among science, pedagogy, and journalism]. *Bakhtiniana, 11*(2), 171–194. https://doi.org/10.1590/2176-457323671

Myers, G. (2003). Discourse studies of scientific popularization: Questioning the boundaries. *Discourse Studies, 5*(2), 265–279. https://doi.org/10.1177/1461445603005002006

Secko, D. M., Amend, E., & Friday, F. (2013). Four models of science journalism. *Journalism Practice, 7*(1), 62–80. https://doi.org/10.1080/1751278 6.2012.691351

Trench, B. (2008). Towards an analytical framework of science communication models. In D. Cheng (Ed.), *Communicating science in social contexts* (pp. 119–135). Springer Science+Business Media.

Van Dijck, J. (2010). Search engines and the production of academic knowledge. *International Journal of Cultural Studies, 13*(6), 574–592. https://doi.org/10.1177/1367877910376582

CHAPTER 3

Methodological Considerations: Frameworks and Rubrics

Abstract This chapter reviews the diverse ways in which popularization discourse is analyzed in the current academic literature. First, it discusses goals and formats of text analysis in general. We specifically focus on quantitative text analysis as a way to produce data matrices and qualitative text analysis to categorize data into themes. In the literature, popularization discourse is analyzed either through frameworks or rubrics. Frameworks give insight into textual components, or strategies, whereas rubrics contain assessment criteria. In this chapter, the main insight is that although current frameworks and rubrics do provide insight into popularization discourse as a genre, it is impossible to produce one overarching framework of strategies that make up popularization discourse purely from these frameworks/rubrics. This gap also points to bigger methodological issues in the current academic literature, which are also discussed in this chapter.

Keywords Quantitative text analysis • Qualitative text analysis • Rubrics • Frameworks • Assessment • Text analysis

3.1 Introduction

This chapter focuses on the analysis of popularization discourse. Before we introduce our own analytical framework in Chap. 4, we take a step back to give an overview of current frameworks for popularization discourse and

© The Author(s) 2023
F. M. Sterk, M. M. van Goch, *Re-presenting Research*,
https://doi.org/10.1007/978-3-031-28174-7_3

to critically discuss their usefulness and applicability. The chapter contains an overview of methodological considerations that are relevant when constructing frameworks, and an overview of the academic literature on current frameworks and rubrics to analyze and assess popularized texts.

3.2   TEXT ANALYSIS: GOALS AND FORMATS

While the previous chapter discussed how scientific data and insights can be communicated to the general public through texts, we will now turn to those texts themselves. In other words, we will look at those popularized texts as research objects or *as generators of academic data and insights.* Often, analytical frameworks are used for this cause, also sometimes called coding schemes. Frameworks and coding schemes ideally provide a way to analyze texts objectively, reliably, and in collaboration with multiple researchers or analysts.

The overarching aim of text analysis is to reconceptualize text as data to be analyzed. In other words, by treating text as a research object, text analysis can provide insight, either qualitative or quantitative, about it. Multiple research traditions have their own forms of text analysis, such as linguistics, computer sciences, and social sciences. In this book, we will focus solely on the linguistic tradition, where the main aim of text analysis is to describe text structure (Roberts, 2000). Broadly seen, there are two types of linguistic text analysis: qualitative (Kuckartz, 2014, 2019) and quantitative (Roberts, 2000). Quantitative text analysis is either representational (researchers classify the *intention* of the writer) or instrumental (researchers apply a theory to interpret the text). Quantitative analysis always produces a data matrix, which can then be used in statistical analysis. Options for quantitative text analysis are thematic text analysis (occurrence of themes), semantic text analysis (relations among themes), and network text analysis (networks of interrelated themes) (Roberts, 2000). Qualitative text analysis, on the other hand, is defined through the use of categories. The aim is to reduce complexity through classification based on characteristics, which can be derived from theory (Kuckartz, 2014). Research that uses qualitative text analysis uses categories (or codes) to develop a category system or coding scheme. Categories can be developed in a construct-driven (deductive) or data-driven (inductive) way, or as a mix of these two options. These categories can be factual, thematic, evaluative, analytical, theoretical, natural, or formal (Kuckartz, 2019). The specific interpretation of what a category is, or how a category can be described or analyzed, often remains implicit—which is a problem that we time and time again encountered during our own research. This sentiment is reflected by Kuckartz:

> The question of what exactly a category represents in empirical research is hardly addressed in literature on research methods, even in textbooks that focus on methods of qualitative data analysis, it is more or less assumed that people already know what a category is, based on common sense. Instead of a definition, you often find a collection of category attributes, particularly in textbooks about qualitative data analysis. There it can be read, for example, that categories should be 'rich', 'meaningful', 'distinguishable', or 'disjunctive'. (Kuckartz, 2014, p. 39)

In the academic literature, presented strategies or categories often lack meaningful descriptions or explanations. Yet analytical frameworks are used because they allow for objective, reliable, and shared analysis. In this book, we have therefore added extensive notes on the strategy (that is, category) level to give a rich description of what each category in the analytical framework means, as such avoiding the pitfall that was discussed in the above quote by Kuckartz. We are proposing a framework that can be used both quantitatively and instrumentally (to score the occurrence of categories) or qualitatively and representationally (to explain the specific use of each category). But before we delve deeper into our framework, let us explore existing frameworks and coding schemes that are used to analyze popularization discourse.

3.3   Text Analysis of Popularization Discourse

In the current literature, popularization discourse is analyzed in multiple ways. Analysis can focus on the achievement of communicative goals (see Metcalfe, 2019), on content analysis (see Kessler, 2019; Shea, 2015), on a specific textual feature (see Rakedzon et al., 2017; Sharon & Baram-Tsabari, 2014 for the analysis of jargon; Riesch, 2015 for the analysis of humor), on componential analysis (August et al., 2020; Giannoni, 2008; Hyland, 2010; Luzón, 2013; Motta-Roth & Lovato, 2009; Nwogu, 1991), or on assessment of popularization discourse in educational settings (Moni et al., 2007; Poronnik & Moni, 2006; Rakedzon & Baram-Tsabari, 2017a, 2017b; Yuen & Sawatdeenarunat, 2020). In this chapter, we focus on two forms of analysis of popularization discourse, as they come closest to providing an overview of the textual features in the genre: componential analysis through frameworks and assessment in educational settings through rubrics. Frameworks can take many different forms, ranging from a list of components to an overview of scoring criteria. Rubrics, on the other hand, often follow a strict template. They are presented in the form of a table, with the assessment criteria (the skills that are graded through the rubric) on one axis and the grades on the other. At each criteria/grade intersection, the table provides an explanation of the assessment point, thus forming a guideline for teachers in their grading (Stevens et al., 2012). The following sections will first cover frameworks that describe popularization strategies, followed by an overview of rubrics that deal with the assessment of popularization discourse.

3.4   FRAMEWORKS FOR POPULARIZATION STRATEGIES

Currently, structural or textual models of popularization discourse are scarce, and an overarching analytical framework—a framework that spans across text types and academic disciplines— is non-existent. Only a handful of researchers have analyzed the use of popularization discourse on the level of structural components, that is, textual strategies or structural categories (August et al., 2020; Giannoni, 2008; Hyland, 2010; Luzón, 2013; Motta-Roth & Lovato, 2009; Nwogu, 1991). We will briefly describe the set-up of each of these studies.

Nwogu (1991) analyzed 15 journalistic reported versions of medical texts to explore their discourse structure. Swales' genre analysis model was used as a theoretical framework. The study resulted in an overview of eight moves and constituent elements. These moves are presented chronologically, that is to say, in a text they always appear in the same order. Giannoni (2008) studied popularization features in 40 journal editorials from medicine and applied linguistics and found seven popularization features. Motta-Roth and Lovato (2009) focused on the rhetorical organization of 30 popularization news articles. They used Nwogu's (1991) framework as a basis and created a new list of six moves that occur in a specific order, combined with two types of discursive elements that can occur throughout the text. Hyland (2010) analyzed the use of proximity in 120 research articles versus popular science articles and found five thematic strategies. Luzón (2013) used an a priori coding scheme from the literature and then employed grounded theory to analyze 75 science blog posts. The research yielded a framework with three themes and 23 strategies. August et al.

(2020) used manual coding of 337 sample articles and computational analysis to analyze the use of ten popularization strategies in a corpus of 128,000 documents.

The structural components, or strategies, that each of these studies delineated can be found in Table 3.1. Because the frameworks from Motta-Roth and Lovato (2009) and Nwogu (1991) work (partly) with a specific *order* of strategies, they are presented specifically as 'moves.' Two other studies worth mentioning are Nisbet et al. (2003) and Calsamiglia and Van Dijk (2004), but because their focus was on a specific topic (stem cells and the genome), the framing is too narrow to compare them with the other sources.

3.5   Discussion of Current Frameworks

In this section we discuss similarities and differences between the frameworks, as well as critically analyzing their construction and the insights they generate. There are multiple similarities between the structural components of the frameworks. All frameworks mention the strategy of including main findings. Analogy/metaphor and describing the method are mentioned in four out of six frameworks; impact/implication and explanation of terms in three; and personalization, question, humor, reader engagement, opinion, and contextualization in two. Dissimilarities between frameworks also exist, with many textual components, such as the addition of a title or inclusive pronouns, being mentioned just once. Furthermore, mismatches are visible between the levels at which components are mentioned, for example as a main-level strategy versus a sub-strategy. Reference to the authors is mentioned as a sub-strategy of presenting new research and describing the data collection procedure in Nwogu's (1991) framework, yet in Motta-Roth and Lovato's (2009) it is part of presentation of the research and voice switching. Another aspect to keep in mind is the order of linguistic moves; only the frameworks by Nwogu (1991) and Motta-Roth and Lovato (2009) presume a specific order.

It thus becomes clear that although these frameworks are connected to the same research problem and show common delineators, there is no real consensus about which strategies or structural components appear in popularization discourse. The studies described are mostly isolated projects— the exception here is the study by Motta Roth and Lovato (2009) that used the Nwogu (1991) study as a point of departure. This also points to

Table 3.1 Overview of popularization strategies mentioned in the literature

Author	Main strategies	Sub-strategies	Lower-level strategies
August et al. (2020, p. 5328)	• Lede • Main (findings) • Impact • Explanation • Analogy • Story • Personal • Jargon • Active • Present		
Giannoni (2008, p. 216)	• Questions • Metaphors • Marked lexis • Humor • Personalization • Appeals to reader • Contingency		
Hyland (2010, pp. 119, 122, 123, 125)	• Organization • Argument structures • Credibility • Stance • Reader engagement		

Luzón (2013, pp. 437, 438)	Rhetorical category	Contextualizing the research
		Announcing the new finding or the new contribution to the discipline
		Describing (and evaluating) method
		Presenting, explaining (and evaluating) results
		• Adopting a neutral or positive stance toward the findings
		• Questioning some aspects of the results
		• Criticizing the whole research and findings
		Drawing implications or highlighting the significance of the study
		• Highlighting the significance of the research for science
		• Broader implications (political, ethical, ideological)
		• Implications for people's lives
		• Implications for involved actors
	Strategies to tailor information	• Explanation of terms and concept
		• Paraphrases/reformulations
		• Comparisons/metaphors
		• Examples from daily life
		• Links
		• Visuals conveying information
	Strategies to engage the reader	• Titles
		• References to popular lore, beliefs
		• Self-disclosure
		• Features of conversational discourse
		• Inclusive pronouns
		• References to reader
		• Questions
		• Humor
		• Positive evaluation
		• Negative evaluation
		• Personal expression of opinion
		• Expressions of feelings or emotional reactions

(continued)

Table 3.1 (continued)

Author	Main strategies	Sub-strategies	Lower-level strategies
Motta-Roth and Lovato (2009, p. 246; translated from Portuguese)	Move 1: Lede/Conclusion of the research (forecast)		
	Move 2: Presentation of the research by	• Identification of researchers (or) • Detailing the results (and) • Reference to the research objective (or) • Allusion to the published scientific article (or the thesis/dissertation)	
	Move 3: Reference to previous knowledge (contextualization) by	• Reference to established knowledge in the area • Emphasis on the social perspective • Allusion to previous research • Indication of limitations in the established knowledge	
	Move 4: Description of methodology by	• Identification of the experimental procedure • Reference to data (source, range, date, location, category)	
	Move 5: Explanation of the results of the search by	Exposure of the results Comparison of current and previous research in terms of:	• Established knowledge • Methodology used • Results obtained
	Move 6: Recommendation of the research conclusions by	• Mention of research implications • Suggestion for future research • Emphasis on the local perspective • Indication of research limitations	
	Voice switching (for more positive or negative comments and opinions) which may include, in addition to the journalist's own voice that underlies all popularization of science news, the voice of:	• The researcher (or metaphorically of the study) • Colleague/technician/institution • Government • Public	
	Explanation of principles and concepts (through rewriting as stance, glosses, and metaphor)		

Nwogu (1991, pp. 115–116)	Move 1: Presenting background information	• By reference to established knowledge in the field • By reference to main research problem • By stressing the local angle • By explaining principles and concepts
	Move 2: Highlighting overall research outcome	• By reference to main research results
	Move 3: Reviewing related research	• By reference to previous research • By reference to limitations of previous research
	Move 4: Presenting new research	• By reference to authors • By reference to research purpose
	Move 5: Indicating consistent observations	• By stating important results • By reference to specific observations
	Move 6: Describing data collection procedure	• By reference to authors • By reference to source of data • By reference to data size
	Move 7: Describing experimental procedure	• By recounting main experimental processes
	Move 8: Explaining research outcome	• By stating a specific outcome • By explaining principles and concepts • By indicating comments and views • By indicating significance of main research outcome • By contrasting present and previous outcomes
	Move 9: Stating research conclusions	• By indicating implications of the research • By promoting further research • By stressing the local angle

a larger issue: the use of popularization strategies does not appear to be a standardized research topic and therefore there is no chronology in reasoning or projects building from one another.

Another important issue with the aforementioned studies has to do with their construction and validity. Datasets are generally (very) small, ranging from 15 to 120 texts, except for the research by August et al. (2020) in which a 128,000-document corpus was used. It should be noted though that this corpus was analyzed using computational analysis, with human coders hand-coding a sample of only 337 articles. This shows the boundaries of human coding in linguistics, with computational analysis offering a chance at coding corpora many times the size of what is usually achievable in hand-coded research. More generally, it is questionable how reliable the frameworks are when they are based on a small dataset, often consisting of texts from a specific subtype of popularization discourse or a single academic field/couple of academic fields. For a framework to be reliable and all-encompassing, it should be based on the analysis of multiple subtypes of popularization and draw sources from multiple disciplinary fields.

Half of the sources do not explain the methodological steps that were taken in the construction of their presented framework. Although Nwogu (1991) mentioned that a genre analysis model was used, no information was given on data analysis or construction of the framework. Likewise, Giannoni (2008) and Hyland (2010) provided little information on methodological steps taken. The other three sources do provide more detail. Luzón (2013) described how grounded theory was used in combination with an a priori framework based on literature. August et al. (2020) showed in detail how writing strategies were chosen, the computational analysis was conducted, and a subset of texts was hand-coded on a sentence level. Motta-Roth and Lovato (2009) explained how three rounds of analysis consisted of individual analysis, cross-analysis, and identification of linguistic components. Because not every source offers an equally clear explanation of the methodological steps that have been taken in the construction, this poses issues for the reproducibility of the research. It also means that it is difficult to estimate the reliability and overall applicability of the frameworks.

Most of these studies are single-author papers, meaning the analysis was also conducted by a single author, or, if multiple people worked on the analysis, this was never mentioned (Giannoni, 2008; Hyland, 2010; Luzón, 2013; Nwogu, 1991). Motta-Roth and Lovato (2009) did use multiple raters and checked for consensus in their analyses, although there is no report on inter-rater reliability. The exception is August et al. (2020) who used Krippendorff's α to measure inter-rater agreement for the manual coding of annotations and reported findings on a strategy level. Ideally, a framework should be used consistently by different raters. By not using multiple raters to construct or evaluate the framework, no knowledge is available about inter-rater reliability, or in other words, the consistency of coding across raters (Holton, 2007; Hallgren, 2012; Kuckartz, 2014). A lack of data on inter-rater reliability hampers the usability of these frameworks, as there is no assurance that the findings that are produced through them are reliable across users.

The biggest issue that surfaces through the analysis of current frameworks is that none of these studies factually presents *a framework for analysis*, in the sense that they report solely on *results* of text analysis. These results are presented as a list of strategies, and in some papers examples are provided. Whether the strategies that are found in one subgenre of popularization or in sources from one discipline can be generalized to other texts or subgenres within popularization discourse remains unclear. In the same vein, these lists do not (or cannot) provide any information on how they should be used in other analytic studies; that is to say, no meta-text, coding information, size of coding, or coding manual is presented in any of these studies. Consequently, there is a lack of validation of strategies in all discussed studies. An exception here is August et al. (2020), who specified their coding to be on the sentence level and reported the accuracy of the computational analysis on a strategy level. Generally speaking, the lack of explicit insight into the methodological choices made in these studies makes it impossible for these lists of strategies to be used as analytical frameworks in follow up studies into popularization discourse.

3.6   Rubrics for Popularization Discourse

Apart from the frameworks discussed above, the structural components of popularization discourse can also be captured in rubrics, which are used in education for assessment purposes. The literature about the assessment of popularization discourse stems from the research field of science communication. Some studies focus on the learning goals that should be implemented in science communication courses (see Baram-Tsabari & Lewenstein, 2013, 2017a, 2017b; Bray et al., 2012; Mercer-Mapstone & Kuchel, 2015). Alternatively, studies focus on the *assessment* of popularization skills acquired in those courses, which is often conducted using rubrics (Moni et al., 2007; Poronnik & Moni, 2006; Rakedzon & Baram-Tsabari, 2017a, 2017b; Yuen & Sawatdeenarunat, 2020). The research field also includes rubrics that primarily consider speaking skills (see Alias & Osman, 2015; Sevian & Gonsalves, 2008; Murdock, 2017), but these were left out of the current overview, to retain the focus on writing skills.

We compared rubrics from five studies. Two studies focused on rubric construction (Rakedzon & Baram-Tsabari, 2017b; Yuen & Sawatdeenarunat, 2020), whereas the other three used a rubric as an assessment tool for learning outcomes of explicit instruction in an educational setting (Moni et al., 2007; Poronnik & Moni, 2006; Rakedzon & Baram-Tsabari, 2017a). Moni et al. (2007) used an opinion editorial rubric to assess learning outcomes for science communication skills in final-year physiology and pharmacology students. Poronnik and Moni (2006) used an opinion editorial rubric for peer review and to assess learning outcomes in undergraduate physiology students. Rakedzon and Baram-Tsabari (2017b) constructed and evaluated a rubric to assess L2 (non-native) STEM graduate students' popularized writing, which was then used in a pretest-posttest intervention study by Rakedzon and Baram-Tsabari (2017a) to assess popularization skills in L2 science and engineering graduate students. Yuen and Sawatdeenarunat (2020) used popular news articles written by science undergraduate students in a rubric development cycle to construct a science communication rubric. Table 3.2 gives an overview of the assessment items mentioned in these rubrics.

3.7   Discussion of Current Rubrics

Some similarities exist in the content of these rubrics. The rubrics by Moni et al. (2007) and Poronnik and Moni (2006) are very similar; these papers are in large part written by the same authors. The same goes for the two identical rubrics by Rakedzon and Baram-Tsabari (2017a, 2017b). Overall, more differences than similarities are visible between rubrics. Each can explain part of the picture of popularization discourse assessment, but the dissimilarities between the rubrics signify that none can give an overarching view (that is, spanning across text forms and academic disciplines). Rubric operationalization is largely the same in all studies; rubrics comprise an assessment grid with scoring options for a range of grades for each assessment criterion from 1 to 8% (Moni et al., 2007), 1–2% to 9–10%

Table 3.2 Overview of assessment items in popularization rubrics

Author	Main strategies	Sub-strategies	Lower-level strategies
Moni et al. (2007, p. 170)	Content	• Key facts and ideas clearly stated • Sufficient background is provided, enables understanding of key ideas • Key ideas are plausible/innovative	
	Genre requirements	• Argument flows in cohesive/logical manner • Argument conforms to structure/length of opinion editorial • Argument addresses needs of the audience • Argument is consistent throughout	
	Quality of writing	• Grammar, syntax, and spelling are of a publishable standard	

(continued)

Table 3.2 (continued)

Author	Main strategies	Sub-strategies	Lower-level strategies
Poronnik and Moni (2006, p. 75)	Content	• Key facts and ideas clearly stated • Sufficient background is provided, enables clear understanding of key ideas • Presentation flows in cohesive/logical manner • Presentation demonstrates broader knowledge	
	Genre requirements	• Conforms to structure/length of opinion editorial • Addresses needs intended audience • Is consistent throughout	
	Quality of writing	• Grammar/syntax/spelling are publishable quality	
Rakedzon and Baram-Tsabari (2017a, Appendix 2/2017b, Appendix E)	• Title • Use of active voice • Inverted pyramid (bottom line then background) format • Journalistic format—six Wh questions (what, who, when, where, why, how) • Definition/explanation • Readability		

Yuen and Sawatdeenarunat (2020, p. 56)	Clarity	Accessibility	• Explanation key finding • Selection of material • Grammar/syntax
		Organization of ideas	• Position of the moves • Macro-organization • Micro-organization
	Color	Significance of key findings	• Implication of key finding for science and the public • Rationale of the study • Significance of key finding
		Language strategies to appeal and engage readers	• Use of appeals to entice readers • Writing style/tone • Use of evaluative language

(Poronnik & Moni, 2006), 1 to 4 (Rakedzon & Baram-Tsabari, 2017a, 2017b), or 1 to 8 (Yuen & Sawatdeenarunat, 2020). Because of these differences in rating scale, the rubrics are not easily comparable. The rubrics by Moni et al. (2007) and Poronnik and Moni (2006) are specifically constructed for one text type, the opinion editorial. This makes them less broadly applicable and more difficult to compare to other rubrics. More generally, and this is true for all rubrics to some degree, the rating of a certain criterion is (partly) dependent upon a subjective assessment by the rater.

Rubric construction and validation are only discussed in three studies. Rakedzon and Baram-Tsabari (2017b) constructed their rubric in a five-stage model that included developing course goals, choosing assessment tasks, setting their standard, developing assessment criteria, and rating values for scoring. Its validation was conducted by checking scoring consistency with two raters. In Rakedzon and Baram-Tsabari (2017a), the rubric was constructed based on course materials and previous research. It was piloted in two rounds, after which empirically developed descriptors were added. Yuen and Sawatdeenarunat (2020) developed quality definitions in their rubric through the analysis of science-related newspaper articles. Student ability, rater severity, item difficulty, and quality of the rating scale were calibrated using a Many-facet Rasch Model. Raters were asked to also mark sample scripts and were interviewed about their marking. A survey was used to measure student perception of the rubric. In these studies, underlying methodologies differ greatly, which hampers the compatibility of rubrics—in other words, it is difficult to construct one overarching insight in assessment criteria. In Moni et al. (2007) and Poronnik and Moni (2006), the construction of the rubric is not discussed explicitly. These rubrics suffer from a methodological gap; it is unclear how sound their construction and how valid their use in practice is.

Just as is the case with the frameworks that were described in the previous two paragraphs, there seems to be no overarching line in academic advancement of rubrics, with researchers working independently and projects not being used for follow-up studies (the exceptions here are the studies conducted by mostly the same authors). This is a pity as it means that there is no clear academic advancement of insights, and each research team that is working on the topic seems to be reinventing a slightly different wheel.

What You Have Learned in This Chapter

- Text analysis sees texts as research objects and turns them into data either through quantitative or qualitative analysis.
- In the current literature, popularization discourse is either analyzed through frameworks or assessed through rubrics. The results from these methods show a whole range of structural components, strategies, or assessment criteria in popularization discourse.
- It is impossible to produce one overarching overview of strategies because of differences in subtypes of popularization discourse and disciplines from which source materials are gathered.
- There is a lack of explanation of methodological steps undertaken in some of these studies, which points to a bigger methodological issue in the research field.
- There is a need for an analytical framework, or in other words, a coding scheme, for popularization discourse.

References

Alias, A., & Osman, K. (2015). Assessing oral communication skills in science: A rubric for development. *Asia Pacific Journal of Educators and Education, 30*, 107–122. Retrieved from http://apjee.usm.my/APJEE_30_2015/APJEE%20 30%20Art%207%20(105%20-%20122).pdf

August, T., Kim, L., Reinecke, K., & Smith, N. A. (2020). Writing strategies for science communication: Data and computational analysis. *Proceedings of the 2020 Conference on Empirical Methods in Natural Language Processing*, 5327–5344. https://doi.org/10.18653/v1/2020.emnlp-main.429

Baram-Tsabari, A., & Lewenstein, B. V. (2013). An instrument for assessing scientists' written skills in public communication of science. *Science Communication, 35*(1), 56–85. https://doi.org/10.1177/1075547012440634

Baram-Tsabari, A., & Lewenstein, B. V. (2017a). Preparing scientists to be science communicators. In P. G. Patrick (Ed.), *Preparing informal science educators: Perspectives from science communication and education* (pp. 437–471). Springer. https://doi.org/10.1007/978-3-319-50398-1_22

Baram-Tsabari, A., & Lewenstein, B. V. (2017b). Science communication training: What are we trying to teach? *International Journal of Science Education, Part B, 7*(3), 285–300. https://doi.org/10.1080/21548455.2017.1303756

Bray, B., France, B., & Gilbert, J. K. (2012). Identifying the essential elements of effective science communication: What do the experts say? *International Journal of Science Education, Part B, 2*(1), 23–41. https://doi.org/10.108 0/21548455.2011.611627

Calsamiglia, H., & Van Dijk, T. A. (2004). Popularization discourse and knowledge about the genome. *Discourse & Society, 15*(4), 369–389. https://doi.org/10.1177/0957926504043705

Giannoni, D. S. (2008). Popularizing features in English journal editorials. *English for Specific Purposes, 27*(2), 212–232. https://doi.org/10.1016/2006.12.001

Hallgren, K. A. (2012). Computing inter-rater reliability for observational data: An overview and tutorial. *Tutorials in Quantitative Methods for Psychology, 8*(1), 23–34. https://doi.org/10.20982/tqmp.08.1.p023

Holton, J. A. (2007). The coding process and its challenges. In A. Bryant & K. Charmaz (Eds.), *The SAGE handbook of grounded theory* (pp. 265–290). SAGE Publications.

Hyland, K. (2010). Constructing proximity: Relating to readers in popular and professional science. *Journal of English for Academic Purposes, 9*(2), 116–127. https://doi.org/10.1016/j.jeap.2010.02.00

Kessler, S. H. (2019). Science communication research in the German-speaking countries: A content analysis of conference abstracts. *Studies in Communication Sciences, 19*(2), 243–251. https://doi.org/10.24434/j.scoms.2019.02.012

Kuckartz, U. (2014). Qualitative text analysis: A guide to methods, practice and using software. *SAGE Publications.* https://doi.org/10.4135/9781446288719

Kuckartz, U. (2019). Qualitative text analysis: A systematic approach. In G. Kaiser & N. Presmeg (Eds.), *Compendium for early career researchers in mathematics education (pp. 181-198).* SpringerOpen. https://doi.org/10.1007/978-3-030-15636-7_8

Luzón, M. J. (2013). Public communication of science in blogs: Recontextualizing scientific discourse for a diversified audience. *Written Communication, 30*(4), 428–457. https://doi.org/10.1177/0741088313493610

Mercer-Mapstone, L. D., & Kuchel, L. J. (2015). Core skills for effective science communication: A teaching resource for undergraduate science education. *International Journal of Science Education, Part B, 7*(2), 181–201. https://doi.org/10.1080/21548455.2015.1113573

Metcalfe, J. (2019). Comparing science communication theory with practice: An assessment and critique using Australian data. *Public Understanding of Science, 28*(4), 382–400. https://doi.org/10.1177/0963662518821022

Moni, R. W., Hryciw, D. H., Poronnik, P., & Moni, K. B. (2007). Using explicit teaching to improve how bioscience students write to the lay public. *Advances in Physiology Education, 31,* 167–175. https://doi.org/10.1152/advan.00111.2006

Motta-Roth, D., & Lovato, C. dos Santos. (2009). Organização retórica do gênero notícia de popularização da ciência: um estudo comparativo entre português e inglês [Rhetorical organization of the science popularization news genre: A comparative study between Portuguese and English]. *Linguagem em (dis)curso, 9*(2), 233–271. https://doi.org/10.1590/S1518-76322009 000200003

Murdock, R. C. (2017). *An instrument for assessing the public communication of scientists*. [Doctoral thesis, Iowa State University]. Iowa State University Digital Repository. https://lib.dr.iastate.edu/etd/15586/

Nisbet, M. C., Brossard, D., & Kroepsch, A. (2003). Framing science: The stem cell controversy in an age of press/politics. *The International Journal of Press/Politics, 8*(2), 36–70. https://doi.org/10.1177/2F1081180X02251047

Nwogu, K. (1991). Structure of science popularizations: A genre analysis approach to the schema of popularized medical texts. *English for Specific Purposes, 10*(2), 111–123. https://doi.org/10.1016/0889-4906(91)90004-G

Poronnik, P., & Moni, R. W. (2006). The opinion editorial: teaching psychology outside the box. *Advances in Psychological Education, 30*(2), 73–82. https://doi.org/10.1152/advan.00075.2005

Rakedzon, T., & Baram-Tsabari, A. (2017a). Assessing and improving L2 graduate students' popular science and academic writing in an academic writing course. *Educational Psychology, 37*(1), 48–66. https://doi.org/10.108 0/01443410.2016.1192108

Rakedzon, T., & Baram-Tsabari, A. (2017b). To make a long story short: A rubric for assessing graduate students' academic and popular science writing skills. *Assessing Writing, 32*, 28–42. https://doi.org/10.1016/j.asw.2016.12.004

Rakedzon, T., Segev, E., Chapnik, N., Yosef, R., & Baram-Tsabari, A. (2017). Automatic jargon identifier for scientists engaging with the public and science communication educators. *PLoS ONE, 12*(8), e0181742. https://doi.org/10.1371/journal.pone.0181742

Riesch, H. (2015). Why did the proton cross the road? Humour and science communication. *Public Understanding of Science, 24*(7), 768–775. https://doi.org/10.1177/0963662514546299

Roberts, C. W. (2000). A conceptual framework for quantitative text analysis: On joining probabilities and substantive inferences about texts. *Quality & Quantity, 34*(3), 259–274. https://doi.org/10.1023/A:1004780007748

Sevian, H., & Gonsalves, L. (2008). Analysing how scientists explain their research: A rubric for measuring the effectiveness of scientific explanations. *International Journal of Science Education, 30*(11), 1441–1467. https://doi.org/10.1080/09500690802267579

Sharon, A. J., & Baram-Tsabari, A. (2014). Measuring mumbo jumbo: A preliminary quantification of the use of jargon in science communication. *Public Understanding of Science, 23*(5), 528–546. https://doi.org/10.1177/096 3662512469916

Shea, N. A. (2015). Examining the nexus of science communication and science education: A content analysis of genetics news articles. *Journal of Research in Science Teaching, 52*(3), 397–409. https://doi.org/10.1002/tea.21193

Stevens, D. D., Levi, A. J., & Walvoord, B. E. (2012). *Introduction to rubrics: An assessment tool to save grading time, convey effective feedback, and promote student learning* (2nd ed.). Stylus Publishing.

Yuen, B. P. L., & Sawatdeenarunat, S. (2020). Applying a rubric development cycle for assessment in higher education: An evidence-based case study of a science communication module. *Asian Journal of the Scholarship of Teaching and Learning, 10*(1), 53–68. Retrieved from: https://nus.edu.sg/cdtl/engagement/publications/ajsotl-home/archive-of-past-issues/V10n1/v10n1-Gan-Sapthaswaran

Construction and Application: Introduction of the Analytical Framework for Popularization Discourse

Abstract In this chapter, we introduce our own analytical framework for popularization discourse. The framework had to comply with four aims: be usable in any subgenre of popularization discourse, be usable in disciplinary and multidisciplinary/interdisciplinary settings, be reliable with multiple raters, and be easy to apply by offering application remarks and explanations of strategies. The analytical framework is produced through a construction and validation step. It consists of 34 strategies that are captured under five themes: Subject Matter, Tailoring Information to the Reader, Credibility, Stance, and Engagement. This chapter offers insight into these themes as well as into each individual strategy, for which application remarks and further reading suggestions are also offered. Lastly, the analytical framework is compared to existing frameworks and rubrics that were proposed in the academic literature.

Keywords Framework • Coding scheme • Themes • Strategies • Text analysis

F. M. Sterk, M. M. van Goch, *Re-presenting Research*,
https://doi.org/10.1007/978-3-031-28174-7_4

4.1   INTRODUCTION

In this chapter we present our analytical framework for popularization discourse. First, we discuss the main aims that our framework should comply with. Then, we offer insight into the development of the framework. The framework consists of five themes, which are explained, and 34 strategies, which are elaborated upon by offering an explanation and application remarks, as well as suggestions for further reading. Lastly, the analytical framework is compared to existing frameworks and rubrics in the academic literature.

4.2   CONSIDERATIONS IN SETTING UP THE FRAMEWORK

In Chap. 3, we discussed existing insights into textual features, or strategies, in popularization discourse. The discussion covered frameworks (August et al., 2020; Giannoni, 2008; Hyland, 2010; Luzón, 2013; Motta-Roth & Lovato, 2009; Nwogu, 1991) as well as rubrics (Moni et al., 2007; Poronnik & Moni, 2006; Rakedzon & Baram-Tsabari, 2017a, 2017b; Yuen & Sawatdeenarunat, 2020). The discussion showed issues in the reliability and usability of these frameworks and rubrics; they do not show a clear line in the strategies that are presented, they are not compatible among each other, they are not validated or constructed with the use of multiple raters in mind, they often consist of results from text analysis and as such do not present a coding scheme, and they mostly cover subgenres of popularization within specific disciplinary settings. Therefore, an overarching framework that is *usable* in the analysis of popularization discourse is still missing from the academic literature and, perhaps more importantly, from practice. Such a framework should ideally comply with four aims. An analytical framework for popularization discourse is:

1. usable in any subgenre of popularization discourse
2. usable in disciplinary but also multidisciplinary and interdisciplinary settings
3. reliable for use with multiple raters
4. easy to apply by offering application remarks and explanations of strategies

4.3 METHODOLOGY

In this section we will present a brief overview of the methodology and set-up of our analytical framework.

The methodology consisted of a construction step and a validation step. The aim of the construction step was to gather a corpus of newspaper articles based on a single academic source text. This would ensure that texts and strategies presented in them are easily comparable to each other and to the academic source material. To achieve this aim, 140 first-year undergraduate liberal education students were asked to write a newspaper article based on the source text "#Sleepyteens: Social media use in adolescence is associated with poor sleep quality, anxiety, depression and low self-esteem" (Woods & Scott, 2016). Participants were asked to read this publication before class. In class, they were asked to write a text about it that would be suitable to publish in the science section of a quality Dutch newspaper. The text had to be within a 400-word limit. The target audience of the newspaper article consisted of a general audience that was interested in science but did not necessarily receive higher education training.

This corpus of newspaper articles was then analyzed for the occurrence of popularization strategies. We worked in test rounds, analyzing 10 randomly selected texts from the corpus in each round. The strategies from Luzón's (2013) research into science blogs were used as a list of a priori codes in the first round, after which descriptive coding was used (Saldaña, 2015) to indicate if each of the strategies was used. Simultaneous coding was allowed, meaning text could be coded for multiple strategies. We used consensual coding (Schmidt, 2004): we compared coding and discussed difficulties and uncertainties. This process of coding and adapting the framework was an iterative process in which, during each round, the a priori code list was adapted according to the insights that were generated through deliberation about the coding, by splitting, merging, deleting, and adding codes. In each new round, the adapted list was used as an a priori list. This process continued until code saturation was reached after six rounds. In a seventh and final round, 10 texts that led to the most discrepancies earlier on were re-analyzed. Throughout the coding rounds, many adaptations were made to the list of codes; some were added or deleted, others split or merged. On the resulting list of 34 codes, pattern coding was used as a second-cycle coding technique to thematize strategies (Saldaña, 2015). Luzón's (2013) framework had originally contained three themes; rhetorical category, strategies to tailor information, and strategies to engage the reader. These themes were reworded into Subject Matter, Tailoring Information to the Reader, and Engagement. Furthermore, the themes Credibility and Stance were introduced—which are also used in Hyland's (2010) framework—to thematize the strategies that fell outside of the scope of the existing three themes.

The inter-rater reliability was checked after each round by using both percent agreement and Cohen's kappa with 95% confidence intervals. Cohen's kappa is used for inter-rater reliability between two raters and controls for agreement because of random guesses. Scores can range between 0 and 1 and include 95% confidence intervals as kappa is an estimate of inter-rater reliability (see McHugh, 2012). In the first round, the inter-rater reliability for our analytical framework was 0.55 (0.46–0.65 confidence intervals), showing a weak level of agreement. In the next rounds, the reliability increased and ultimately reached a kappa of 0.90 (0.86–0.95 confidence intervals) in the seventh and final round, denoting an almost perfect level of agreement.

The aim of the validation step was to check the framework against a corpus of texts from multiple different subgenres of popularization discourse, written by professionals, and containing multiple topics and source texts. This validation step was needed because all texts in the construction round were based on a single source text from a single academic field, thus creating the possibility that some strategies could not be employed. Furthermore, the corpus in the construction phase was written by students, not professional writers, thus creating the possibility that some strategies were not used. This second corpus consisted of 38 science journalism articles written by a range of professional media outlets that were chosen according to Berezow's (2017) infographic on quality of science reporting. The two axes of this infographic present the compellingness of the content and the degree of evidence-based reporting of different media outlets. This creates a three-by-three grid ranging from 'evidence-based reporting' with 'almost always compelling science content,' to 'ideologically driven reporting/poor reporting' with 'no compelling science content.' The corpus contained texts from all represented quadrants of science reporting (for example, the option 'always compelling science content' with 'ideologically driven/poor reporting' was not represented by any outlets), apart from the quadrant of 'not usually compelling science content' with 'ideologically driven/poor reporting,' as this is seen as poor science reporting overall. The corpus contained many different topics (and thus disciplinary fields) and multiple subtypes of popularization such as news articles and overview articles. The reader can find more about this corpus in Chap. 6. The corpus was checked against the analytical framework with the particular aim of analyzing how often strategies were used and to check if any previously undiscovered strategies could be found. In this round, one strategy was deleted and five additional strategies were added to the already constructed themes. The final framework contains 34 strategies captured under five themes and will be explained in the next two sections.

4.4 THEMES

In this section we describe the five themes that comprise the framework: Subject Matter, Tailoring Information to the Reader, Credibility, Stance, and Engagement.

Subject Matter includes strategies concerning content from the original scientific publication. Although it is impossible to construct one single organizational structure of popularization texts, several rhetorical strategies usually appear in them (Luzón, 2013). Instead of a "narrative of science," popularization texts provide a "narrative of nature" (Hyland, 2010, pp. 120–121). By moving the main claim to the first paragraph, the focus shifts to novelty and importance. The object studied becomes more important than the methodological steps taken (Hyland, 2010). Uncertainties that would be discussed in the academic text are removed and the focus is on results (Fahnestock, 1986).

Tailoring Information to the Reader contains recontextualization strategies that remodel academic findings to an everyday-life and understandable setting. In academic texts a shared base of knowledge is assumed between writer and reader, yet this is not the case in popularizations (Hyland, 2010). Therefore, popularizations need to be recontextualized from the academic context to that of the lay audience and need to be perceived as suitable within the new context (Gotti, 2014). Information is tailored to readers by connecting to what they (are presumed to) already know, through explanations or connections to everyday life (Hyland, 2010). In part, this recontextualization takes place by focusing on the application and consequences of a phenomenon (Fahnestock, 1986).

Credibility is created in academic texts when authors position themselves in relation to other researchers and publications. For popularization texts, it is assumed that readers do not possess disciplinary knowledge or cross-disciplinary expertise, so credibility of the source is emphasized instead. Researchers important to the topic under discussion are mentioned and credibility is constructed through their academic position. Furthermore, quotes underline the credibility of the presented material. In academic texts, credibility increases through depersonalization as it suggests objectivity. The opposite is true for popularized texts, in which personalization strategies are used (Hyland, 2010).

In popularization discourse, *Stance* on the topic under discussion can also be communicated. The media contribute to opinions that are formed about research and researchers (Calsamiglia & Van Dijk, 2004). Furthermore, personal attitude and the expression of stance play a big part in constructing proximity. In academic texts, hedges are used to indicate that researchers are careful in their statements, but these are removed in popularized texts to create more impact for academic findings. Instead, the popularization writer uses stance and opinions to comment on the research or the publication to engage the reader (Hyland, 2010).

Engagement is used to connect to readers; writers use it to signal their awareness of the audience's presence. Engagement strategies use discourse that is informal and geared toward the reader to get their attention, create a shared understanding, include them as participants in the discourse, and influence them (Hyland, 2010; Luzón, 2013). Many strategies that are part of this theme play an active role in reformulation.

4.5   Strategies

The five themes together contain 34 strategies. Table 4.1 provides an overview of each strategy in the analytical framework. Each strategy contains an explanation that is based on literature and application remarks that are based on our experience from working with the framework. The 'further reading' column presents additional sources that cover each strategy—the interested reader can explore each strategy further using these sources.

4.6   The Framework versus the Literature

We established that, ideally, an analytical framework should be compliant with four aims. Our framework is usable in any subgenre of popularization because in its construction and validation phase, multiple subforms of popularization were taken into consideration. The framework can be used in any disciplinary, multidisciplinary, or interdisciplinary field because texts from the validation phase represented a range of disciplinary fields, none of which led to any issues in coding, and the student writing was produced as part of an interdisciplinary undergraduate program. Furthermore, there is no disciplinary bias in any of the themes or strategies in the framework. The framework is reliably usable by multiple raters because we worked with multiple raters during the construction phase and this collaboration produced a reliable kappa. In fact, we would advise anyone working with this framework to analyze texts in duos and thoroughly discuss differences of opinion, as this process is very insightful for learning how the analytical framework works, as well as generating insight into your own frame of reference. Lastly, the framework is easy to use because we offer application remarks and an explanation of each strategy.

As was mentioned in Chap. 3, it is difficult to make one single overview of all popularization strategies from the available frameworks. Still, it is possible to relate the framework that we just presented to the existing literature. First, the connection to previous sources can be seen in the 'further reading' column in Table 4.1, which shows other sources in which a particular strategy is also presented (it should be noted though that not all sources mentioned in this column present an explicit *framework*). Except for LINK TO THE ACADEMIC PUBLICATION, all the strategies in our framework are mentioned in at least one of these sources. Some features are mentioned in multiple frameworks, such as main findings (in our framework:

Table 4.1 The framework of analysis for popularization discourse

Theme	Strategy	Explanation	Application remarks	Further reading
Subject Matter	LEDE	A LEDE is a short, introductory section used to establish the most important findings. Its goal is to attract the attention of the reader.	• Often typographically distinct, for example presented in boldface. • Only code when presented between the title and first paragraph of the main body. • Usually used only once. • Length is between one and a couple of sentences.	August et al. (2020) and Motta-Roth and Lovato (2009)
	CONTEXTUALIZATION	CONTEXTUALIZATION is an organizational strategy used to introduce the research or the topic. In contextualization, a scenario is created that is supposed to attract the attention of the reader and present the topic in an everyday-life context. The main claim can be presented at the end of the contextualization as a lead-up to newsworthiness.	• Constructed through strategies such as NOVELTY, EXAMPLES FROM DAILY LIFE, QUESTIONS, and IMAGERY. Also mark these strategies when used in CONTEXTUALIZATION. • CONTEXTUALIZATION is possible without ANNOUNCING THE NEW FINDING OR NEW CONTRIBUTION TO THE DISCIPLINE. • Often in the first paragraph of the text. • Usually used only once. • Length is between a sentence and a paragraph.	Gotti (2014), Hyland (2010), and Luzón (2013)
	ANNOUNCING THE NEW FINDING OR NEW CONTRIBUTION TO THE DISCIPLINE	ANNOUNCEMENTS focus on the news value or 'newness' of research. They are used to underpin validation for the production of a text. For readers, ANNOUNCEMENTS underpin a text's readability. ANNOUNCING THE NEW FINDING OR NEW CONTRIBUTION TO THE DISCIPLINE consists of an announcement claim that states research has been conducted and a newsworthiness claim that focuses on the new results.	• Code even when only the announcement claim or only the newsworthiness claim is presented. • Often presented in the first or second paragraph. • Used only once in single-source popularizations, used multiple times in overview texts about a topic. • Length is between one and a couple of sentences.	Hyland (2010) and Luzón (2013)
	NOVELTY	NOVELTY shows the motivation for doing research by giving an overview of previously known knowledge, showing a knowledge gap, or pointing out why research is necessary. Popularization texts use NOVELTY to achieve newsworthiness.	• Code even when only an overview of previously known knowledge or a gap in knowledge or the necessity of research is presented. • Often presented in the first or second paragraph. • Usually used only once. • Length is between one and a couple of sentences.	Hyland (2010), Molek Kozakowska (2017), Motta-Roth and Lovato (2009), and Nwogu (1991)
	DESCRIBING THE METHOD	The METHOD is often presented in non-technical terms, so that it appeals to readers with different levels of disciplinary knowledge. METHODS can be described either on an abstract and design-level (hypothesis/goal/topic) or a practical/applied level (measures and methods used).	• Code any remark (either abstract or concrete) about the METHODS used or about the design of the study. • Can be presented in any paragraph, usually not in the first or last. • Length is between a couple of words and a couple of sentences.	Hyland (2010), Luzón (2013), Motta-Roth and Lovato (2009), and Nwogu (1991)
	PRESENTING RESULTS/CONCLUSIONS	PRESENTING RESULTS/CONCLUSIONS entails showing the new information and conclusions from academic research. They are presented as new insights and sometimes explained further.	• An often-used strategy and a strategy that often takes up a large percentage of the text. • Be aware that as part of PRESENTING RESULTS/CONCLUSIONS, many other strategies can appear that should also be coded. • Be aware that the boundary between PRESENTING RESULTS/CONCLUSIONS and showing APPLIED IMPLICATIONS can be vague. • Similarly, PRESENTING RESULTS/CONCLUSIONS can be part of other strategies, such as TITLE, LEDE, and ANNOUNCING THE NEW FINDING OR NEW CONTRIBUTION TO THE DISCIPLINE. • Can be presented in any paragraph, often all paragraphs between CONTEXTUALIZATION and the final paragraph.	August et al. (2020), Hyland (2010), Luzón (2013), Motta-Roth and Lovato (2009), and Nwogu (1991)

Tailoring Information to the Reader	Applied Implications	APPLIED IMPLICATIONS recontextualize knowledge beyond the scope of research, to show their application in everyday life and everyday terms. APPLIED IMPLICATIONS can include ethical, cultural, ideological, and political implications. In some cases, readers are urged to take action.	• Be aware that ACADEMIC IMPLICATIONS or implications for actors in the academic world are another strategy as part of Credibility. Only implications beyond the scope of research should be coded as APPLIED IMPLICATIONS. • When readers are urged to take action, usually they are referenced through REFERENCES TO THE READER or INCLUSIVE PRONOUNS, and an imperative verb is used. • Usually presented in the final paragraph. • Usually used only once. • Length is between one and a couple of sentences.	August et al. (2020), Calsamiglia (2003), Giannoni (2008), Gotti (2014), Luzón (2013), Molek-Kozakowska (2017), Motta-Roth and Lovato (2009), and Nwogu (1991)
	Explanations	EXPLANATIONS are used to elaborate upon a term or idea. They consist of paraphrases, reformulations in which specialist discourse is presented into more understandable language, explanations of definitions, or elaborations of terms and concepts.	• Code any type of explanatory statement. • Disciplinary or specialist discourse from the original publication is at the core of this strategy. When presented verbatim, also code for DIRECT QUOTE FROM THE ACADEMIC PUBLICATION. • Presented in any part of the text. • Length is between a few words and a paragraph.	August et al. (2020), Gotti (2014), Hyland (2007, 2010), Luzón (2013), Motta-Roth and Lovato (2009), and Nwogu (1991)
	Imagery	IMAGERY is used to help readers integrate new knowledge; it lets readers relate new knowledge to existing knowledge. IMAGERY consists of all types of explanatory elements that use figurative language such as metaphor, analogy, comparison, and idioms.	• Code any type of IMAGERY. • Presented in any part of the text. • Length is between a few words and a paragraph. Analogy and metaphor are usually not longer than a couple of words.	August et al. (2020), Calsamiglia and Van Dijk (2004), Giannoni (2008), Luzón (2013), and Motta-Roth and Lovato (2009)
	Examples from daily life	EXAMPLES FROM DAILY LIFE create a scenario to draw information into an everyday context, in order to explain it. This can include definitions, explanations, facts, and concepts as long as they are part of the everyday example.	• Often accompanied by REFERENCES TO THE READER, INCLUSIVE PRONOUNS, and FEATURES OF CONVERSATIONAL DISCOURSE. • This strategy excludes SELF-DISCLOSURE OF THE AUTHOR'S PUBLIC OR PERSONAL LIFE; this is coded as a separate strategy. • Presented in any part of the text. • Length is between a few words and a couple of sentences.	Gotti (2014), Hyland (2010), and Luzón (2013)
	Hyperlinks	HYPERLINKS link to online sources that contain explanations and additional information, or to related stories.	• Typographically distinct; usually printed in blue and/or underlined. • Medium dependent; will be non-clickable without an internet connection, will not be used in print media. • Presented in any part of the text. • Length is between one and a few words.	Luzón (2009, 2013)
	Visuals	VISUALS attract attention or are used as visual explanatory elements. They can be presented in the form of pictures, graphics, and diagrams.	• VISUALS can include captions and credits; also code these as part of the visual. The caption might contain other strategies (for example, PRESENTING RESULTS/CONCLUSIONS); code these separately. • Medium dependent: might not be used in non-digitized texts. • Presented in any part of the text. Pictures are often used to attract attention and presented at the top of the text.	Hyland (2010) and Luzón (2013)

(continued)

Table 4.1 (continued)

Theme	Strategy	Explanation	Application remarks	Further reading
Credibility	ACADEMIC IMPLICATIONS	ACADEMIC IMPLICATIONS present the implications for actors involved in disseminating and publishing research. The implications might already be presented as part of the original academic text. In popularizations they are used to underpin the quality of the research.	• Be aware that implications beyond the scope of research, for example ethical, cultural, ideological, and political implications, are a different strategy and should be coded as APPLIED IMPLICATIONS. • Usually presented toward the end of the text. • Length is between one and a couple of sentences.	Luzón (2013) and Nwogu (1991)
	MENTIONING MORE RESEARCH IS NECESSARY/NEXT STEP IN RESEARCH	MENTIONING MORE RESEARCH IS NECESSARY/NEXT STEP IN RESEARCH consists of two parts: a signaling mention that more research is necessary, and an argumentative part about what the content of the research/the next step in research should be.	• Code even when only the signaling claim or only the argumentative claim about content is presented. • Presented toward the end of the text. • Usually used only once. • Length is between one and a couple of sentences.	Motta-Roth and Lovato (2009) and Nwogu (1991)
	CONTRIBUTION TO RESEARCH	The CONTRIBUTION TO RESEARCH is presented as part of the transformation of academic findings into newsworthy claims. New information is presented as a scientific breakthrough by highlighting the significance of the results for the scientific community and for the development of the academic field.	• Be aware that newsworthiness is also constructed in terms of the contribution to everyday life, but these are separate strategies (see APPLIED IMPLICATIONS and EXAMPLES FROM DAILY LIFE). • Presented toward the end of the text. • Usually used only once. • Length is between one and a couple of sentences.	Hyland (2010) and Nwogu (1991)
	MENTION OF STATISTICS	STATISTICS in popularizations are based on the statistical analysis in academic papers but presented in a more comprehensible way. They are often simple statistical terms (means, percentages). STATISTICS are used to underpin the credibility of the research and give insight into the research results.	• Percentages can be presented either as a percent mark ('33%') or in words ('a third'). • Presented in any part of the text. • Length of a single word or two words.	Hijmans et al. (2003) and Hyland (2010)
	GIVING THE RESEARCHER AN ACTIVE VOICE/DIRECT QUOTES FROM THE RESEARCHER	GIVING THE RESEARCHER AN ACTIVE VOICE is especially important in constructing credibility. Because new information cannot be embedded within a discipline and its knowledge base, credibility is constructed through explanations from researchers themselves.	• Often, but not always, presented in quotation marks. • Often introduced through a marker such as 'the researcher says....' • Be aware that quotes from anyone other than the researchers are coded as GIVING NON-RESEARCHERS AN ACTIVE VOICE. • Presented in any part of the text. • Length is between a couple of words and a couple of sentences.	Gotti (2014), Hyland (2010), Motta-Roth and Lovato (2009), and Nwogu (1991)
	LEXICAL MENTION OF THE ORIGINAL RESEARCH	LEXICAL MENTIONS are used to construct credibility of the source, for example through the position of the researcher in an academic institution.	• LEXICAL MENTIONS can include researchers' names, researchers' academic positions, university, title of the research, and journal where the research is published. • Presented in any part of the text. • Length is between a single word and a couple of words.	Hyland (2010), Luzón (2013), Motta-Roth and Lovato (2009), and Nwogu (1991)

Group	Element	Definition	Operationalization	References
	ADDITIONAL SOURCES	ADDITIONAL SOURCES are used to add information or underpin findings from a different perspective. They are the popularized counterpart of the way sources are cited in academic texts to position new claims versus already established knowledge.	• Sometimes accompanied by a HYPERLINK. • Presented in any part of the text. • Length is between one and a couple of sentences.	Hijmans et al. (2003) and Hyland (2010)
	LINK TO THE ACADEMIC PUBLICATION	A LINK TO THE ACADEMIC PUBLICATION is either presented as an in-text hyperlink or in a separate sentence. The credibility is constructed through showing the original source of the new information.	• When a hyperlink is used to LINK TO THE ACADEMIC PUBLICATION, code under HYPERLINK too. • Usually presented either in the first paragraph as an in-text hyperlink or at the end of the text in a separate sentence. • Medium dependent; will be non-clickable without an internet connection, will not be used in print media. • Usually used only once. • Length is between one word and a sentence.	Luzón (2009)
	DIRECT QUOTE FROM THE ACADEMIC PUBLICATION	A DIRECT QUOTE FROM THE ACADEMIC PUBLICATION consists of part of the academic text that is re-presented as a quote.	• The fact that the quote is lifted from the academic publication is explicitly stated. • Presented in quotation marks. • Presented in any part of the text. • Length is between a couple of words and a couple of sentences.	Nwogu (1991)
	IN-TEXT SPECIFICATION OF A SOURCE	IN-TEXT SPECIFICATION OF A SOURCE consists of an explicit remark about the origins of information.	• Linguistic markers include '… has told the publication through email…' or 'stated in a new press release….' • Presented in any part of the text though always connected to GIVING THE RESEARCHER AN ACTIVE VOICE or GIVING THE NON-RESEARCHER AN ACTIVE VOICE. • Length is a couple of words.	Nwogu (1991)
Stance	OPINION	OPINIONS consist of any evaluative remark. They can be positive or negative, for example to praise research, to express criticism about research, or to highlight its importance.	• Code any evaluative remark, whether positive or negative. • Any participant in the discourse can contribute an OPINION; this means that OPINIONS in quotes also count. • Presented in any part of the text. • Length is between one and a couple of sentences.	Luzón (2013), Molek Kozakowska (2017), and Motta-Roth and Lovato (2009)
	STANCE MARKERS	STANCE MARKERS are used to convey attitudes, emotions, and evaluations. Epistemic stance comments on certainty, doubt, reliability, limitations, and the reality or actuality of a proposition. Attitudinal stance is used to convey attitudes, feelings, or value judgments. Style stance comments on the way information is being given or should be understood.	• STANCE MARKERS might be difficult to analyze as their interpretation is also dependent upon a shared understanding between the writer and reader. Take the context and target audience into consideration. • All types of stance markers are captured under one strategy: STANCE MARKERS. • STANCE MARKERS are often used as part of OPINIONS. • Presented in any part of the text. • Usually used multiple times. • Length is one word or sometimes two words.	Aull and Lancaster (2014), Biber (2006), Conrad and Biber (1999), Giannoni (2008), and Gray and Biber (2014)

(continued)

Table 4.1 (continued)

Theme	Strategy	Explanation	Application remarks	Further reading
Engagement	TITLES/SUBHEADINGS	TITLES and SUBHEADINGS attract the attention of the reader, for example by presenting (part of) the main claim of the academic text or connecting results to everyday life.	• Often typographically distinct from the rest of the text, such as in a bigger font or boldface. • Strategies used in TITLES/SUBHEADINGS are marked separately. • TITLES are presented at the top of the text; SUBHEADINGS are presented in any part of the text. • TITLE used once, SUBHEADINGS used multiple times. • Length is between a couple of words and one sentence.	Luzón (2013)
	REFERENCES TO POPULAR LORE AND BELIEFS, AND POPULAR CULTURE	REFERENCES TO POPULAR LORE AND BELIEFS, AND POPULAR CULTURE work by creating a familiar mental landscape of existing understanding of topics in which a connection can be made to new information from research.	• Might be dependent upon the cultural context and frame of reference of the analyst. Take the context and target audience into consideration. • Presented in any part of the text. • Length is between a couple of words and a couple of sentences.	Gamson 1988 as cited in Lievrouw (1990), Luzón (2013), and Zehr (2014)
	SELF-DISCLOSURE OF THE AUTHOR'S PUBLIC OR PERSONAL LIFE	SELF-DISCLOSURE OF THE AUTHOR'S PUBLIC OR PERSONAL LIFE is used to add details about the author's personal or professional life. The defining element of this strategy is the personal connection. SELF-DISCLOSURE OF THE AUTHOR'S PUBLIC OR PERSONAL LIFE consists of examples of the daily and personal life of the author.	• Be aware that the examples must be connected to the personal life of the author. Otherwise, code under EXAMPLES FROM DAILY LIFE. • Presented in any part of the text. • Length is between a couple of words and a couple of sentences.	August et al. (2020), Giannoni (2008), and Luzón (2013)
	INCLUSIVE PRONOUNS	INCLUSIVE PRONOUNS create a shared group between reader and writer in which both parties share the same point of view, or a taken-for-granted view is presented. Note that the term 'inclusive pronouns' is often used to refer to gender neutral or gender inclusive pronouns, but in this strategy INCLUSIVE PRONOUNS refers to plural first-person pronouns (we, us, ours) that are applied to recognize the reader and the writer as one group.	• Plural first-person pronouns. • Be aware that in science communication, authors might write from the plural first-person perspective. In this case, the pronouns are not inclusive as they only refer to the researchers. • Presented in any part of the text. • Usually used multiple times. • Length is a single word.	Hyland (2010) and Luzón (2013)
	REFERENCES TO THE READER	REFERENCES TO THE READER are second-person pronouns used to represent the readers as actors in the interaction. They are employed to acknowledge the readers' presence and draw them into the discourse.	• Singular/plural second-person pronouns. • Presented in any part of the text. • Usually used multiple times. • Length is a single word.	Hyland (2010), Luzón (2013), and Molek-Kozakowska (2017)
	GIVING THE NON-RESEARCHER AN ACTIVE VOICE/DIRECT QUOTES	Where GIVING THE RESEARCHER AN ACTIVE VOICE is employed for credibility, GIVING THE NON-RESEARCHER AN ACTIVE VOICE constructs engagement. Quotes from non-academic experts, journalists, and people with personal experience about a topic are used to include multiple perspectives, to make a connection to the perspective from everyday life, and to gear the text toward different audiences.	• Often, but not always, presented in quotation marks. • Often introduced through a marker such as 'person X says…'. • Be aware that quotes from researchers are coded as GIVING THE RESEARCHER AN ACTIVE VOICE. • Presented in any part of the text. • Length is between a couple of words and a couple of sentences.	Botelho et al. (2016) and Motta-Roth and Lovato (2009)

FEATURES OF CONVERSATIONAL DISCOURSE	FEATURES OF CONVERSATIONAL DISCOURSE consist of any type of everyday-life language use, and are used to give the feeling of informality.	• INCLUSIVE PRONOUNS, REFERENCES TO THE READER, QUESTIONS, and STANCE MARKERS may also be coded as FEATURES OF CONVERSATIONAL DISCOURSE when they contribute to the feeling of colloquial language. • Presented in any part of the text. • Usually used multiple times. • Length is between a single word and multiple paragraphs (the latter in the case of informal language use).	Luzón (2013) and Molek-Kozakowska (2017)
QUESTIONS	QUESTIONS are used to catch the attention of the reader, to create dialogic involvement, to make readers think about a topic, as an organizational tool, to challenge existing views, and as an explanatory tool.	• Typographically distinct through the use of a question mark. • QUESTIONS are often used as part of other strategies, such as TITLES/SUBHEADINGS. • Presented in any part of the text. • Length is usually one sentence.	Giannoni (2008), Hyland (2010), Luzón (2013), and Nwogu (1991)
HUMOR	HUMOR is used to build solidarity with the reader, to reinforce assumptions, and to entertain. It can include light teasing, irony, and sarcasm.	• Might be dependent upon the cultural context and frame of reference of the analyst. Take the context and target audience into consideration. • Presented in any part of the text. • Length is between a single word and a couple of sentences.	Giannoni (2008), Luzón (2013), and Riesch (2015)
EXPLICIT SELF-REFERENCE	EXPLICIT SELF-REFERENCE is used to make a connection with the reader, to express the authority of the writer, or to let a writer express their identity. The writer makes their presence known within the text.	• Singular and plural first-person pronouns. • Be aware that singular and plural first-person pronouns might also be used to refer to the researcher(s) or other actors in the text; count only as EXPLICIT SELF-REFERENCE if they refer to the author of the popularization. • Presented in any part of the text. • Length is a single word.	Biber (2006), Giannoni (2008), Hyland (2002), and Luzón (2013)

PRESENTING RESULTS/CONCLUSIONS), analogy/metaphor (in our framework: IMAGERY), DESCRIBING THE METHOD, APPLIED IMPLICATIONS, EXPLANATIONS, QUESTIONS, HUMOR, OPINION, and CONTEXTUALIZATION. Other features are mentioned once, such as HYPERLINKS, ADDITIONAL SOURCES, and GIVING THE RESEARCHER AN ACTIVE VOICE. On the other hand, our framework is not simply an aggregation of these sources, and not all strategies mentioned in them have become part of the analytical framework.

Furthermore, there are several key differences between earlier frameworks and the framework presented in this book. Examples of strategies that are mentioned in other frameworks that are not part of our framework are story (storytelling/narrative) and active (active voice) (August et al., 2020), contingency (the effect of personal experiences on work or beliefs) (Giannoni, 2008), argument structures (Hyland, 2010), and expressions of feelings or emotional reactions (Luzón, 2013). Because our aim was to construct a framework that was evidence-based and these strategies were not found in our text analyses, and because we did not want to make an aggregation of all strategies that were discussed in the literature, they are not included in our framework. Another difference is that some of these sources might present strategies on a different level, for example as sub-strategies or aggregated into one. This led to issues of incompatibility between existing frameworks. Our framework does not contain sub-strategies, which ensures all strategies are presented on the same level of importance. Furthermore, our framework does not presume a specific order of linguistic moves, like Motta-Roth and Lovato's (2009) and Nwogu's (1991) frameworks do, as we want to enable coders to make inferences based upon their own analyses. However, we do recognize that in the data we used for our research, some moves are often seen in a specific order (CONTEXTUALIZATION, NOVELTY, ANNOUNCING THE NEW FINDING OR NEW CONTRIBUTION TO THE DISCIPLINE), or appear together (EXAMPLES FROM DAILY LIFE with INCLUSIVE PRONOUNS or REFERENCES TO THE READER).

Similarly, none of the popularization rubrics discussed in Chap. 3 can give an overarching insight into popularization discourse, but some of the strategies (or assessment criteria) mentioned in them are also presented in our framework. They are key facts (in our framework: PRESENTING RESULTS/CONCLUSIONS), background (in our framework: CONTEXTUALIZATION, NOVELTY) (Moni et al., 2007; Poronnik & Moni, 2006), TITLES (Rakedzon & Baram-Tsabari, 2017a, 2017b), ACADEMIC IMPLICATIONS and APPLIED IMPLICATIONS (Yuen & Sawatdeenarunat, 2020), and

EXPLANATIONS (Rakedzon & Baram-Tsabari, 2017a, 2017b; Yuen & Sawatdeenarunat, 2020).

These rubrics also explain processes that are needed to construct effective popularization texts, such as the use of active voice (Rakedzon & Baram-Tsabari, 2017a, 2017b) or the use of language strategies to appeal to and engage readers (Yuen & Sawatdeenarunat, 2020). These processes are more about actions a writer should take that ultimately will lead to certain textual elements than they are about those textual elements directly, which is why they are not part of our framework.

In Chap. 2 we discussed the two main textual processes that construct popularization discourse, and through them form popularization strategies: recontextualization and reformulation (Bondi et al., 2013; Calsamiglia & Van Dijk, 2004; Ciapuscio, 2003; Gotti, 2014). Recontextualization entails moving scientific facts from the expert context to the layperson context and, in doing so, presenting specialized knowledge in a way that non-specialized readers can understand (Bondi et al., 2013; Calsamiglia & Van Dijk, 2004). Many strategies in our framework are a form of recontextualization: LEDE, CONTEXTUALIZATION, ANNOUNCING THE NEW FINDING OR NEW CONTRIBUTION TO THE DISCIPLINE, NOVELTY, DESCRIBING THE METHOD, PRESENTING RESULTS/CONCLUSIONS, APPLIED IMPLICATIONS, HYPERLINKS, VISUALS, ACADEMIC IMPLICATIONS, MENTIONING MORE RESEARCH IS NECESSARY/NEXT STEP IN RESEARCH, CONTRIBUTION TO RESEARCH, MENTION OF STATISTICS, LEXICAL MENTION OF THE ORIGINAL RESEARCH, ADDITIONAL SOURCES, LINK TO THE ACADEMIC PUBLICATION, DIRECT QUOTE FROM THE ACADEMIC PUBLICATION, IN-TEXT SPECIFICATION OF A SOURCE, OPINION, TITLES/SUBHEADINGS, REFERENCES TO POPULAR LORE AND BELIEFS, AND POPULAR CULTURE, SELF-DISCLOSURE OF THE AUTHOR'S PUBLIC OR PERSONAL LIFE, and HUMOR. Reformulation remodels the language that is used to the new target audience (Gotti, 2014). Reformulation occurs in the strategies EXPLANATIONS, IMAGERY, STANCE MARKERS, INCLUSIVE PRONOUNS, REFERENCES TO THE READER, FEATURES OF CONVERSATIONAL DISCOURSE, QUESTIONS, and EXPLICIT SELF-REFERENCE. Some strategies are a combination of reformulation and recontextualization elements: GIVING THE RESEARCHER AN ACTIVE VOICE, GIVING THE NON-RESEARCHER AN ACTIVE VOICE, and EXAMPLES FROM DAILY LIFE. These strategies recontextualize the content of the academic text, but they also rephrase it into colloquial language. Our framework also shows that it is possible to construct other goals or themes through recontextualization and reformulation processes. An example here is the theme Credibility, which is entirely made up of recontextualization strategies.

What You Have Learned in This Chapter

- In this chapter, an analytical framework for popularization discourse is presented that is usable in any subgenre of popularization discourse, usable in disciplinary or multidisciplinary/interdisciplinary settings, workable and reliable with multiple raters, and easy to apply.
- The framework consists of five themes (Subject Matter, Tailoring Information to the Reader, Credibility, Stance, and Engagement) and 34 strategies that are explained and supported by further readings and application remarks.
- There are points of overlap between this analytical framework and existing frameworks from the literature, but also key differences; this framework is not an aggregation of existing efforts and aims to improve upon existing work.

References

August, T., Kim, L., Reinecke, K., & Smith, N. A. (2020). Writing strategies for science communication: Data and computational analysis. *Proceedings of the 2020 Conference on Empirical Methods in Natural Language Processing*, 5327–5344. https://doi.org/10.18653/v1/2020.emnlp-main.429

Aull, L. L., & Lancaster, Z. (2014). Linguistic markers of stance in early and advanced academic writing: A corpus-based comparison. *Written Communication, 31*(2), 151–183. https://doi.org/10.1177/074108831 4527055

Berezow, A. (2017, March 5). Infographic: The best and worst science news sites. *American Council on Science and Health..* https://www.acsh.org/ news/2017/03/05/infographic-best-and-worst-science-news-sites-10948

Biber, D. (2006). Stance in spoken and written university registers. *Journal of English for Academic Purposes, 5*(2), 97–116. https://doi.org/10.1016/j. jeap.2006.05.001

Bondi, M., Cacchiani, S., & Mazzi, D. (2013). *Discourse in and through the media: Recontextualizing and reconceptualizing expert discourse.* In M. Bondi, S. Cacchiani, & D. Mazzi (Eds.), *Discourse in and through the media: Recontextualizing and reconceptualizing expert discourse* (pp. 1–21). Cambridge Scholars Publishing.

Botelho, J. S., Martins, S., & Coura-Sobrinho, J. (2016). Autonimic modulation in science dissemination: A look into the work of journalists from Folha de S. Paulo Online and from some international news agencies. *Bakhtiniana, 11*(2), 14–32. https://doi.org/10.1590/2176-457323538

Calsamiglia, H. (2003). Popularization discourse. *Discourse Studies, 5*(2), 139–146. https://doi.org/10.1177/1461445603005002307

Calsamiglia, H., & Van Dijk, T. A. (2004). Popularization discourse and knowledge about the genome. *Discourse & Society, 15*(4), 369–389. https://doi.org/10.1177/0957926504043705

Ciapuscio, G. E. (2003). Formulation and reformulation procedures in verbal interactions between experts and (semi-)laypersons. *Discourse Studies, 5*(2), 207–233. https://doi.org/10.1177/1461445603005002004

Conrad, S., & Biber, D. (1999). Adverbial marking of stance in speech and writing. In S. Hunston & G. Thompson (Eds.), *Evaluation in text: Authorial stance and the construction of discourse* (pp. 56–73). Oxford University Press.

Fahnestock, J. (1986). Accommodating science: The rhetorical life of scientific facts. *Written Communication, 3*(3), 275–296. https://doi.org/10.117 7/2F0741088386003003001

Giannoni, D. S. (2008). Popularizing features in English journal editorials. *English for Specific Purposes, 27*(2), 212–232. https://doi.org/10.1016/2006.12.001

Gotti, M. (2014). Reformulation and recontextualization in popularization discourse. *Ibérica, 27*, 15–34.

Gray, B., & Biber, D. (2014). Stance markers. In K. Aijmer & C. Rühlemann (Eds.), *Corpus pragmatics: A handbook* (pp. 219–248). Cambridge University Press. https://doi.org/10.1017/CBO9781139057493.012

Hijmans, E., Pleijter, A., & Wester, F. (2003). Covering scientific research in Dutch newspapers. *Science Communication, 25*(2), 153–176. https://doi.org/10.1177/1075547003259559

Hyland, K. (2002). Authority and invisibility: Authorial identity in academic writing. *Journal of Pragmatics, 34*(8), 1091–1112. https://doi.org/10.1016/S0378-2166(02)00035-8

Hyland, K. (2007). Applying a gloss: Exemplifying and reformulating in academic discourse. *Applied Linguistics, 28*(2), 266–285. https://doi.org/10.1093/applin/amm011

Hyland, K. (2010). Constructing proximity: Relating to readers in popular and professional science. *Journal of English for Academic Purposes, 9*(2), 116–127. https://doi.org/10.1016/j.jeap.2010.02.00

Lievrouw, L. A. (1990). Communication and the social representation of scientific knowledge. *Critical Studies in Mass Communication, 7*(1), 1–10. https://doi.org/10.1080/15295039009360159

Luzón, M. J. (2009). Scholarly hyperwriting: The function of links in academic weblogs. *Journal of the American Society for Information Science and Technology, 60*(1), 75–89. https://doi.org/10.1002/asi.20937

Luzón, M. J. (2013). Public communication of science in blogs: Recontextualizing scientific discourse for a diversified audience. *Written Communication, 30*(4), 428–457. https://doi.org/10.1177/0741088313493610

McHugh, M. L. (2012). Interrater reliability: The kappa statistic. *Biochemia Medica, 22*(3), 276–282. https://doi.org/10.11613/BM.2012.031

Molek-Kozakowska, K. (2017). Communicating environmental science beyond academia: Stylistic patterns of newsworthiness in popular science journalism. *Discourse & Communication, 11*(1), 69–88. https://doi.org/10.1177/1750 481316683294

Moni, R. W., Hryciw, D. H., Poronnik, P., & Moni, K. B. (2007). Using explicit teaching to improve how bioscience students write to the lay public. *Advances in Physiology Education, 31*, 167–175. https://doi.org/10.1152/advan.00111.2006

Motta-Roth, D., & Lovato, C. dos Santos. (2009). Organização retórica do gênero notícia de popularização da ciência: um estudo comparativo entre português e inglês [Rhetorical organization of the science popularization news genre: A comparative study between Portuguese and English]. *Linguagem em (dis)curso, 9*(2), 233–271. https://doi.org/10.1590/S1518-763220090 0200003

Nwogu, K. (1991). Structure of science popularizations: A genre analysis approach to the schema of popularized medical texts. *English for Specific Purposes, 10*(2), 111–123. https://doi.org/10.1016/0889-4906(91)90004-G

Poronnik, P., & Moni, R. W. (2006). The opinion editorial: teaching psychology outside the box. *Advances in Psychological Education, 30*(2), 73–82. https://doi.org/10.1152/advan.00075.2005

Rakedzon, T., & Baram-Tsabari, A. (2017a). Assessing and improving L2 graduate students' popular science and academic writing in an academic writing course. *Educational Psychology, 37*(1), 48–66. https://doi.org/10.108 0/01443410.2016.1192108

Rakedzon, T., & Baram-Tsabari, A. (2017b). To make a long story short: A rubric for assessing graduate students' academic and popular science writing skills. *Assessing Writing, 32*, 28–42. https://doi.org/10.1016/j.asw.2016.12.004

Riesch, H. (2015). Why did the proton cross the road? Humour and science communication. *Public Understanding of Science, 24*(7), 768–775. https://doi.org/10.1177/0963662514546299

Saldaña, J. (2015). *The coding manual for qualitative researchers* (3rd ed.). SAGE Publications.

Schmidt, C. (2004). The analysis of semi-structured interviews. In U. Flick, E. von Kardoff, & I. Steinke (Eds.), *A companion to qualitative research* (pp. 253–258). SAGE Publications.

Woods, H. C., & Scott, H. (2016). #Sleepyteens: Social media use in adolescence is associated with poor sleep quality, anxiety, depression and low self-esteem. *Journal of Adolescence, 51,* 41–49. https://doi.org/10.1016/j.adolescence.2016.05.008

Yuen, B. P. L., & Sawatdeenarunat, S. (2020). Applying a rubric development cycle for assessment in higher education: An evidence-based case study of a science communication module. *Asian Journal of the Scholarship of Teaching and Learning, 10*(1), 53–68. Retrieved from: https://nus.edu.sg/cdtl/engagement/publications/ajsotl-home/archive-of-past-issues/V10n1/v10n1-Gan-Sapthaswaran

Zehr, E. P. (2014). Avengers Assemble! Using pop-culture icons to communicate science. *Advances in Physiology Education, 38*(2), 118–123. https://doi.org/10.1152/advan.00146.201

Text Example: Using an Analytical Framework to Code a Professional Science Journalism Text

Abstract In this chapter, the analytical framework for popularization discourse is used to code the science journalism text "Baby Poop Is Loaded With Microplastics." The chapter shows the kinds of insights that can be produced on the level of the individual text. Furthermore, presenting the in-depth analysis of a single text allows us to share what coding using our framework looks like on the level of the strategy.

Keywords Science journalism • Text analysis • Text example • Analytical framework • Strategies

5.1 Introduction

In this chapter we present an example of using our framework as an analytic tool. One professional science journalism text is analyzed using the framework. The goal of this analysis is to assess which strategies a professional science journalist uses, at the textual level, to communicate research

© The Author(s) 2023
F. M. Sterk, M. M. van Goch, *Re-presenting Research*,
https://doi.org/10.1007/978-3-031-28174-7_5

findings to the general public. Thus, the question driving the analysis was which of the 34 strategies in our framework were discernible in the professional text. Therefore, this analysis is quantitative and instrumental. The goal of the chapter is to show what results to expect when using the framework to analyze one individual text at the strategy level.

5.2   Corpus Construction

This chapter presents the analysis of a single text: "Baby Poop Is Loaded With Microplastics," written by Matt Simon and published on the website of Wired in September 2021 (Simon, 2021). Wired is a journalism platform (including a magazine and website) that focuses on the impact of new technology on everyday life. The text details new published research that shows microplastics have been found in newborn babies' first feces; it offers information about the research as well as drawing those insights into the bigger realm of everyday life to show what this discovery about microplastics might mean for us on a daily and individual level. The text was part of the corpus of science journalism texts that was used in the validation step of the construction of our framework (see Chaps. 4 and 6). We chose the text because it uses the most strategies out of any text in the corpus of professional science journalism writing: 27 out of 34. Here, we show the outcomes of using the framework on the level of a single text.

5.3   Analysis

In this section, we show the analysis of the text "Baby Poop is Loaded With Microplastics". We have underlined and coded each strategy. Some codes are used more than once, others not at all. To signal codes that run beyond a single line of text, we added a raised edge to the underline. Codes that run beyond multiple lines of text are coded with a vertical line. The underlining is color coded per theme: Subject matter is coded in purple, Tailoring Information to the Reader in blue, Credibility in green, Stance in orange, and Engagement in pink.

Baby Poop Is Loaded With Microplastics

imagery   title

An alarming new study finds that infant feces contain 10 times more

stance   opinion

polyethylene terephthalate (aka polyester) than an adult's.

explanation   results

lede +

announcement

visual

Photograph: David Gee/Alamy Stock Photo

Whenever a plastic bag or bottle degrades, it breaks into ever smaller pieces that

work their way into nooks in the environment. When you wash synthetic fabrics,

imagery   ref. reader

tiny plastic fibers break loose and flow out to sea. When you drive, plastic bits fly

example daily life   URL + additional source   ref. reader

off your tires and brakes. That's why literally everywhere scientists look, they're

ref. reader URL + add. source   stance

contextualization

finding microplastics—specks of synthetic material that measure less than 5

conversational discourse   explanation

millimeters long. They're on the most remote mountaintops and in the deepest

URL + add. source   URL

oceans. They're blowing vast distances in the wind to sully once pristine regions

add. source   imagery   imagery

like the Arctic. In 11 protected areas in the western US, the equivalent of 120

URL + add. source

million ground-up plastic bottles are falling out of the sky each year.

URL + add. source   imagery

And now, microplastics are coming out of babies. In a pilot study published today,

stance   conv. discourse + humor   link to academic pub. + URL

scientists describe sifting through infants' dirty diapers and finding an average of

imagery   method   announcement

36,000 nanograms of polyethylene terephthalate (PET) per gram of feces, 10

explanation

times the amount they found in adult feces. They even found it in newborns' first

results

feces. PET is an extremely common polymer that's known as polyester when it's

explanation

used in clothing, and it is also used to make plastic bottles. The finding comes a

year after another team of researchers calculated that preparing hot formula in

plastic bottles severely erodes the material, which could dose babies with several   novelty

stance   stance

million microplastics a day, and perhaps nearly a billion particles a year.

URL + add. source   stance

Although adults are bigger, scientists think that in some ways infants have more

exposure. In addition to drinking from bottles, babies could be ingesting

stance

microplastics in a dizzying number of ways. They have a habit of putting

conversational discourse

everything in their mouths—plastic toys of all kinds, but they'll also chew on

fabrics. (Microplastics that shed from synthetic textiles are known more

explanation

specifically as microfibers, but they're plastic all the same.) Babies' foods are

URL + add. source

wrapped in single-use plastics. Children drink from plastic sippy cups and eat off

contextualization

plastic plates. The carpets they crawl on are often made of polyester. Even

hardwood floors are coated in polymers that shed microplastics. Any of this

imagery

could generate tiny particles that children breathe or swallow.

stance

Indoor dust is also emerging as a major route of microplastic exposure, especially

imagery

for infants. (In general, indoor air is absolutely lousy with them; each year you

humor   ref. reader

could be inhaling tens of thousands of particles.) Several studies of indoor spaces

stance   URL + add. source   URL + add. source

have shown that each day in a typical household, 10,000 microfibers might land on a single square meter of floor, having flown off of clothing, couches, and bed

sheets. Infants spend a significant amount of their time crawling through the

conversational discourse

stuff, agitating the settled fibers and kicking them up into the air.

imagery

example daily life

"Unfortunately, with the modern lifestyle, babies are exposed to so many

active voice researcher

different things for which we don't know what kind of effect they can have later

inclusive pronouns

in their life," says Kurunthachalam Kannan, an environmental health scientist at

lexical reference

New York University School of Medicine and coauthor of the new paper, which

lexical reference

appears in the journal Environmental Science and Technology Letters.

URL + link to academic publication   lexical reference

results

The researchers did their tally by collecting dirty diapers from six 1-year-olds and

imagery

running the feces through a filter to collect the microplastics. They did the same

with three samples of meconium—a newborn's first feces—and stool samples

explanation

from 10 adults. In addition to analyzing the samples for PET, they also looked for

polycarbonate plastic, which is used as a lightweight alternative to glass, for

method

instance in eyeglass lenses. To make sure that they only counted the

microplastics that came from the infants' guts, and not from their diapers, they

ruled out the plastic that the diapers were made of: polypropylene, a polymer

explanation

that's distinct from polycarbonate and PET.

All told, PET concentrations were 10 times higher in infants than in adults, while

conversational discourse

polycarbonate levels were more even between the two groups. The researchers

results

found smaller amounts of both polymers in the meconium, suggesting that

stance

babies are born with plastics already in their systems. This echoes previous

imagery

novelty

studies that have found microplastics in human placentas and meconium.

URL + additional source

What this all means for human health—and, more urgently, for infant health—

contribution to research

scientists are now racing to find out. Different varieties of plastic can contain any

novelty

scientific impl. imagery

of at least 10,000 different chemicals, a quarter of which are of concern for

statistics

people, according to a recent study from researchers at ETH Zürich in

URL + additional source

Switzerland. These additives serve all kinds of plastic-making purposes, like

imagery

providing flexibility, extra strength, or protection from UV bombardment, which

imagery

contextualization

degrades the material. Microplastics may contain heavy metals like lead, but they

stance

also tend to accumulate heavy metals and other pollutants as they tumble

imagery

through the environment. They also readily grow a microbial community of

stance

viruses, bacteria, and fungi, many of which are human pathogens.

imagery   *URL + additional source*

Of particular concern are a class of chemicals called endocrine-disrupting

chemicals, or EDCs, which disrupt hormones and have been connected to

contextualization

reproductive, neurological, and metabolic problems, for instance increased

obesity. The infamous plastic ingredient bisphenol A, or BPA, is one such EDC

conversational discourse   *explanation*

that has been linked to various cancers.

URL + additional source

"We should be concerned because the EDCs in microplastics have been shown to

inclusive pronoun   *active voice researchers*

novelty

be linked with several adverse outcomes in human and animal studies," says Jodi

lex. ref.

Flaws, a reproductive toxicologist at the University of Illinois at Urbana-

lexical reference

Champaign, who led a 2020 study from the Endocrine Society on plastics. (She

URL + additional source

wasn't involved in this new research.) "Some of the microplastics contain

lexical reference   active voice researchers

chemicals that can interfere with the normal function of the endocrine system."

novelty

Infants are especially vulnerable to EDCs, since the development of their bodies

depends on a healthy endocrine system. "I strongly believe that these chemicals

active voice researcher   opinion

do affect early life stages," says Kannan. "That's a vulnerable period."

lexical reference   active voice researcher

novelty

This new research adds to a growing body of evidence that babies are highly

stance   stance

exposed to microplastic. "This is a very interesting paper with some very worrying

stance active voice researcher stance

numbers," says University of Strathclyde microplastic researcher Deonie Allen,

opinion   lexical reference   lexical reference

who wasn't involved in the study. "We need to look at everything a child is

lexical reference   active voice researcher

exposed to, not just their bottles and toys."

scientific implication

contribution to research

Since infants are passing microplastics in their feces, that means the gut could be

stance

absorbing some of the particles, like it would absorb nutrients from food. This is

explanation

known as translocation: Particularly small particles might pass through the gut

stance

results

wall and end up in other organs, including the brain. Researchers have actually

demonstrated this in carp by feeding them plastic particles, which translocated

URL + additional source

through the gut and worked their way to the head, where they caused brain

imagery

novelty

damage that manifested as behavioral problems: Compared to control fish, the

individuals with plastic particles in their brains were less active and ate more

slowly.

But that was done with very high concentrations of particles, and in an entirely

different species. While scientists know that EDCs are bad news, they don't yet

know what level of microplastic exposure it would take to cause problems in the

next step in research

human body. "We need many more studies to confirm the doses and types of

active voice researcher

chemicals in microplastics that lead to adverse outcomes," says Flaws.

lexical reference

In the meantime, microplastics researchers say you can limit children's contact

`reference to the reader`

with particles. Do not prepare infant formula with hot water in a plastic bottle—

use a glass bottle and transfer it over to the plastic one once the liquid reaches

`applied`

`implications`

room temperature. Vacuum and sweep to keep floors clear of microfibers. Avoid

plastic wrappers and containers when possible. Microplastics have contaminated

every aspect of our lives, so while you'll never get rid of them, you can at least

`inclusive pronoun    reference reader        reference reader`

reduce your family's exposure.

`reference reader`

5.4   OVERVIEW OF STRATEGIES

Through the analysis of the text "Baby Poop Is Loaded With Microplastics," it becomes clear that 27 out of 34 possible strategies were used in this text. In Table 5.1, examples of the use of each strategy are provided.

5.5   INTERPRETATION

After using the framework to analyze the text "Baby Poop Is Loaded With Microplastics," the following insights can be drawn. Our inter-rater reliability was 0.84 with 95% confidence intervals of 0.63 to 1.00.

Overall, it could be said that the text adheres to the genre demands of science journalism. The focus of the text is mainly on the results of the research and on how these new insights impact our daily life on an individual level. The use of popularization strategies is dense, with all text in the article coded for at least one strategy and often for multiple strategies. This also means that a lot of overlap is visible in strategy use. Some strategies are coded for longer passages of text, often for a length of multiple sentences, such as CONTEXTUALIZATION, PRESENTING RESULTS/CONCLUSIONS,

Table 5.1 Overview of strategies in "Baby Poop Is Loaded With Microplastics"

Theme	Strategy	Examples of strategy use
Subject Matter	LEDE	An alarming new study finds that infant feces contain 10 times more polyethylene terephthalate (aka polyester) than an adult's.
	CONTEXTUALIZATION	Whenever a plastic bag or bottle degrades, it breaks into ever smaller pieces that work their way into nooks in the environment. When you wash synthetic fabrics, tiny plastic fibers break loose and flow out to sea. When you drive, plastic bits fly off your tires and brakes. That's why literally everywhere scientists look, they're finding microplastics—specks of synthetic material that measure less than 5 millimeters long. They're on the most remote mountaintops and in the deepest oceans. They're blowing vast distances in the wind to sully once pristine regions like the Arctic. In 11 protected areas in the western US, the equivalent of 120 million ground-up plastic bottles are falling out of the sky each year. And now, microplastics are coming out of babies.
	ANNOUNCING THE NEW FINDING OR NEW CONTRIBUTION TO THE DISCIPLINE	An alarming new study finds that infant feces contain 10 times more polyethylene terephthalate (aka polyester) than an adult's.
	NOVELTY	The finding comes a year after another team of researchers calculated that preparing hot formula in plastic bottles severely erodes the material, which could dose babies with several million microplastics a day, and perhaps nearly a billion particles a year.
	DESCRIBING THE METHOD	The researchers did their tally by collecting dirty diapers from six 1-year-olds and running the feces through a filter to collect the microplastics. They did the same with three samples of meconium—a newborn's first feces—and stool samples from 10 adults. In addition to analyzing the samples for PET, they also looked for polycarbonate plastic, which is used as a lightweight alternative to glass, for instance in eyeglass lenses. To make sure that they only counted the microplastics that came from the infants' guts, and not from their diapers, they ruled out the plastic that the diapers were made of: polypropylene, a polymer that's distinct from polycarbonate and PET.
	PRESENTING RESULTS/CONCLUSIONS	All told, PET concentrations were 10 times higher in infants than in adults, while polycarbonate levels were more even between the two groups. The researchers found smaller amounts of both polymers in the meconium, suggesting that babies are born with plastics already in their systems.

Tailoring Information to the Reader	APPLIED IMPLICATIONS	In the meantime, microplastics researchers say you can limit children's contact with particles. Do not prepare infant formula with hot water in a plastic bottle—use a glass bottle and transfer it over to the plastic one once the liquid reaches room temperature. Vacuum and sweep to keep floors clear of microfibers. Avoid plastic wrappers and containers when possible. Microplastics have contaminated every aspect of our lives, so while you'll never get rid of them, you can at least reduce your family's exposure.
	EXPLANATIONS	… they ruled out the plastic that the diapers were made of: polypropylene, a polymer that's distinct from polycarbonate and PET.
	IMAGERY	Indoor dust is also emerging as a major route of microplastic exposure, especially for infants.
	EXAMPLES FROM DAILY LIFE	When you drive, plastic bits fly off your tires and brakes.
	HYPERLINKS	When you wash synthetic fabrics, tiny plastic fibers break loose and flow out to sea. When you drive, plastic bits fly off your tires and brakes.
	VISUALS	 David Gee / Alamy Stock Photo

(continued)

Table 5.1 (continued)

Theme	Strategy	Examples of strategy use
Credibility	ACADEMIC IMPLICATIONS	"We need to look at everything a child is exposed to, not just their bottles and toys."
	MENTIONING MORE RESEARCH IS NECESSARY/NEXT STEP IN RESEARCH	But that was done with very high concentrations of particles, and in an entirely different species. While scientists know that EDCs are bad news, they don't yet know what level of microplastic exposure it would take to cause problems in the human body. "We need many more studies to confirm the doses and types of chemicals in microplastics that lead to adverse outcomes," says Flaws.
	CONTRIBUTION TO RESEARCH	This new research adds to a growing body of evidence that babies are highly exposed to microplastic.
	MENTION OF STATISTICS	Different varieties of plastic can contain any of at least 10,000 different chemicals, a quarter of which are of concern for people, according to a recent study from researchers at ETH Zürich in Switzerland.
	GIVING THE RESEARCHER AN ACTIVE VOICE/DIRECT QUOTES FROM THE RESEARCHER	"Unfortunately, with the modern lifestyle, babies are exposed to so many different things for which we don't know what kind of effect they can have later in their life," says Kurunthachalam Kannan…
	LEXICAL MENTION OF THE ORIGINAL RESEARCH	… says Kurunthachalam Kannan, an environmental health scientist at New York University School of Medicine and coauthor of the new paper, which appears in the journal *Environmental Science and Technology Letters.*
	ADDITIONAL SOURCES	This echoes previous studies that have found microplastics in human placentas and meconium.
	LINK TO THE ACADEMIC PUBLICATION	In a pilot study published today, scientists describe sifting through infants' dirty diapers and finding an average of 36,000 nanograms of polyethylene terephthalate (PET) per gram of feces, 10 times the amount they found in adult feces.
	DIRECT QUOTE FROM THE ACADEMIC PUBLICATION	N/A
	IN-TEXT SPECIFICATION OF A SOURCE	N/A

Stance	Opinion	Stance markers	"I strongly believe that these chemicals do affect early life stages," says Kannan. "That's a vulnerable period." This new research adds to a growing body of evidence that babies are highly exposed to microplastic. "This is a very interesting paper with some very worrying numbers."
Engagement		Titles/subheadings	Baby Poop Is Loaded With Microplastics
		References to popular lore and beliefs, and popular culture	N/A
		Self-disclosure of the author's public or personal life	N/A
		Inclusive pronouns	Microplastics have contaminated every aspect of our lives, so while you'll never get rid of them, you can at least reduce your family's exposure.
		References to the reader	Microplastics have contaminated every aspect of our lives, so while you'll never get rid of them, you can at least reduce your family's exposure.
		Giving the non-researcher an active voice/direct quotes from the non-researcher	N/A
		Features of conversational discourse	Infants spend a significant amount of their time crawling through the stuff, agitating the settled fibers and kicking them up into the air.
		Questions	N/A
		Humor	(In general, indoor air is absolutely lousy with them; each year you could be inhaling tens of thousands of particles.)
		Explicit self-reference	N/A

NOVELTY, DESCRIBING THE METHOD, CONTRIBUTION TO RESEARCH, APPLIED IMPLICATIONS, and MENTIONING MORE RESEARCH IS NECESSARY—these strategies are from the themes Subject Matter, Tailoring Information to the Reader, and Credibility. These strategies never overlap with one another, but rather alternate. Alternatively, strategies that are coded for shorter passages of text, on the level of individual or multiple words, often appear on top of or together with one of the bigger codes. These are STANCE MARKERS, EXPLANATIONS, EXAMPLES FROM DAILY LIFE, REFERENCES TO THE READER, FEATURES OF CONVERSATIONAL DISCOURSE, IMAGERY, HYPERLINKS, ADDITIONAL SOURCES, HUMOR, LEXICAL MENTION OF THE ORIGINAL RESEARCH, LINK TO THE ACADEMIC PUBLICATION, MENTION OF STATISTICS, GIVING THE RESEARCHER AN ACTIVE VOICE, GIVING THE NON-RESEARCHER AN ACTIVE VOICE, OPINION, and INCLUSIVE PRONOUNS. These strategies are mainly from the themes Stance and Engagement. Some strategies are employed throughout the entire text, such as STANCE MARKERS, INCLUSIVE PRONOUNS, REFERENCES TO THE READER, and IMAGERY, while others such as ANNOUNCING THE NEW FINDING OR NEW CONTRIBUTION TO THE DISCIPLINE (start), CONTEXTUALIZATION (start), APPLIED IMPLICATIONS (end), and MENTIONING MORE RESEARCH IS NECESSARY (end) only appear in specific spots in the text. APPLIED IMPLICATIONS are used twice in this text in two different locations, and NOVELTY is used multiple times toward the end of the text instead of at the start, which is something we have not seen happen often in other texts we analyzed.

This example also shows that text analysis includes interpretation to some degree. The text includes the sentence "We need to look at everything a child is exposed to, not just their bottles and toys," a quote from a researcher who was not involved in the study under discussion. This sentence can be construed as an ACADEMIC IMPLICATION, that is, a call to action for researchers to further investigate exposure to microplastics. Alternatively, this sentence can be interpreted to be an APPLIED IMPLICATION, that is, a call to action for the audience to pay more attention to all sources of (micro)plastic that their children are exposed to. In most cases, if a statement is unclear, the meaning can still be gleaned from the context—which in this case has led us to code the statement as an ACADEMIC IMPLICATION—but this is not always a possibility. The choice to code as an ACADEMIC IMPLICATION has consequences for further coding too; 'we' now includes the quoted researcher and other researchers, but not the public, and can therefore not be coded as an INCLUSIVE PRONOUN. Had the statement been coded as an APPLIED IMPLICATION, 'we' would have referred to the researcher and readers, and therefore would be coded as an INCLUSIVE PRONOUN.

This text also shows two uses of 'strategies' which we had not seen before in our corpus analysis, both of which are contained in the following sentence: "'I strongly believe that these chemicals do affect early life stages,' says Kannan." This example includes a quote by a researcher who is not involved in the research under discussion—which is a textual feature that does not really fit under either GIVING THE RESEARCHER AN ACTIVE VOICE or GIVING THE NON-RESEARCHER AN ACTIVE VOICE. The second textual feature is the use of reference within a quote; in this case it is not EXPLICIT SELF-REFERENCE to the writer, but to another actor and voice within the text. These two discoveries show that a framework is never truly finished (see also Chap. 8).

What You Have Learned in This Chapter

- In this chapter an example analysis of one science journalism text, "Baby Poop Is Loaded With Microplastics," was presented.
- The chapter provides insights into what the application of the analytical framework looks like on the level of codes that are produced for a single text.
- On the level of coding, a significant insight that is produced is that some strategies are coded on the level of multiple sentences, and hardly ever overlap one another, while other strategies are coded on the single-word or multi-word level and are overlapping over longer codes.
- Some strategies are employed throughout the entire text, while others are only found in specific spots. Some strategies (NOVELTY, APPLIED IMPLICATIONS) are used differently to what might be expected following the analytical framework.
- The use of quotes from other researchers that are not involved in the study and the use of self-reference within a quote does not comply with any strategy currently in the analytical framework.

REFERENCE

Simon, M. (2021, September 22). Baby poop is loaded with microplastics. *Wired*. https://www.wired.com/story/baby-poop-is-loaded-with-microplastics/

Corpus Example: Using an Analytical Framework to Explore Professional Writing in Science Journalism

Abstract This chapter presents the analysis of a corpus of 38 professional science journalism articles. The analysis is quantitative, that is, it shows how often each strategy is used in the corpus, and instrumental, which means that an interpretation of the texts is offered through the application of the analytical framework for popularization discourse. In this chapter, we include the percentage of use and an example of the use of each strategy. The median of strategy use is 20 (out of a possible 34), with a high score of 27 strategies and a low score of 10 strategies. Some strategies are used in (nearly) every text, such as PRESENTING RESULTS/CONCLUSIONS or HYPERLINKS, whereas other strategies are used hardly at all, such as REFERENCES TO POPULAR LORE AND BELIEFS, AND POPULAR CULTURE. Overall, strategies in the themes Subject Matter, Tailoring Information to the Reader, and Stance are used most, with the aggregated scores for the themes Credibility and Engagement being much lower. This analysis provides insights into how popularization strategies are used in a corpus of science journalism writing as a whole.

Keywords Science journalism • Corpus • Theme • Strategy • Quantitative text analysis

© The Author(s) 2023   83
F. M. Sterk, M. M. van Goch, *Re-presenting Research*,
https://doi.org/10.1007/978-3-031-28174-7_6

6.1   INTRODUCTION

This chapter presents the results of using the analytical framework for popularization discourse to analyze a corpus of professional science journalisms texts. The analysis is quantitative and instrumental. It investigates how often each strategy is used in the texts, aiming to provide insights into the broader genre of science journalism (instead of at the level of the individual text, such as in Chap. 5), to answer questions about how professional science journalists communicate research findings to lay audiences. The goal of the chapter is to show what kinds of results are generated when analyzing a corpus of texts using the framework in this manner. The chapter also includes examples from the professional texts for each strategy.

6.2   CORPUS CONSTRUCTION

This corpus consists of 38 science journalism texts. These texts are written by professional science journalists and published online. The texts cover a range of topics, from immunotherapy in llamas and bioengineered catnip to digital dementia and COVID-19 vaccines, and a range of disciplinary fields, such as biology, archeology, musicology, and technology. They also present a range of popularization subtypes: some articles presented insights from one academic publication, while others took multiple studies into consideration, showed an overview of current knowledge, or presented science news. This corpus was constructed for the validation phase of the development of our framework, using Berezow's (Berezow, 2017) infographic on the quality of science reporting (see Chap. 4). This infographic arranges science reporting on the axes of evidence-based reporting and compellingness of content. For this corpus, texts were used from outlets from all represented quadrants except the quadrant that scores poor on both criteria.

We visited the websites of each of the outlets and downloaded the most recent article from the science section, or if there was none, the general website. In some cases, the most recent article was not an article about research findings, but, for example, an interview. In that case, we downloaded the five most recent articles and chose the one that was most compliant with the genre of science journalism. All texts in the corpus are written by science journalists, bar one text about holograms, which is written by the researchers themselves and therefore technically seen as a form of science communication (see Dahiya, 2021).

6.3   ANALYSIS

We analyzed this corpus using our analytical framework. For each text, we performed an analysis such as the one that was shown in Chap. 5. We then tallied scores for each of the 34 strategies and for each text. In Fig. 6.1, we present a quantitative overview of the number of strategies that are used in each text. The figure shows that although all texts used a variety of strategies, none use (near) all of them.

In Table 6.1, for all 34 strategies, we present the percentage of texts using each strategy. We also show, for all five themes, an aggregated percentage for the number of strategies that are used per theme. In Table 6.2 we present one or multiple examples for the use of each strategy. We furthermore included an overview of additional strategies that are also used in the example shown (since, as discussed in Chap. 5, multiple strategies can be used within one phrase or sentence), to illustrate the complexity of the texts in the corpus and, therefore, the complexity of their analysis. If context is added for the example to make sense, the strategy under discussion is underlined. HYPERLINKS are displayed in blue.

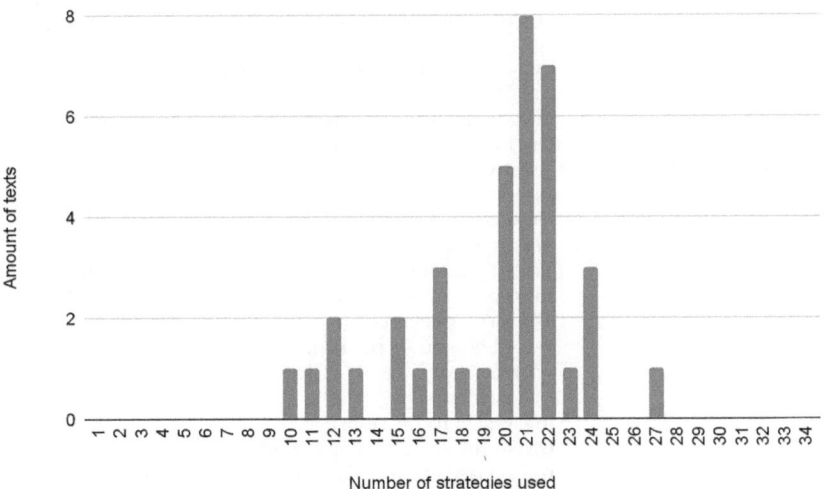

Fig. 6.1 Number of strategies (out of 34 strategies) used per text in a corpus of professional science journalism texts

Table 6.1   Quantitative results from the corpus analysis of professional science journalism texts

Theme	Aggregated percentage of strategies	Strategy	Percentage of texts using the strategy
Subject Matter	75.9%	LEDE	65.8%
		CONTEXTUALIZATION	81.6%
		ANNOUNCING THE NEW FINDING OR NEW CONTRIBUTION TO THE DISCIPLINE	89.5%
		NOVELTY	47.4%
		DESCRIBING THE METHOD	71.1%
		PRESENTING RESULTS/CONCLUSIONS	100%
Tailoring Information to the Reader	80.3%	APPLIED IMPLICATIONS	71.1%
		EXPLANATIONS	71.1%
		IMAGERY	92.1%
		EXAMPLES FROM DAILY LIFE	47.4%
		HYPERLINKS	100%
		VISUALS	100%
Credibility	42.8%	ACADEMIC IMPLICATIONS	28.9%
		MENTIONING MORE RESEARCH IS NECESSARY/NEXT STEP IN RESEARCH	42.1%
		CONTRIBUTION TO RESEARCH	39.5%
		MENTION OF STATISTICS	42.1%
		GIVING THE RESEARCHER AN ACTIVE VOICE/DIRECT QUOTES FROM THE RESEARCHER	84.2%
		LEXICAL MENTION OF THE ORIGINAL RESEARCH	78.9%
		ADDITIONAL SOURCES	71.1%
		LINK TO THE ACADEMIC PUBLICATION	60.5%
		DIRECT QUOTE FROM THE ACADEMIC PUBLICATION	7.9%
		IN-TEXT SPECIFICATION OF A SOURCE	15.8%
Stance	81.6%	OPINION	63.2%
		STANCE MARKERS	100%
Engagement	36.6%	TITLES/SUBHEADINGS	100%
		REFERENCES TO POPULAR LORE AND BELIEFS, AND POPULAR CULTURE	7.9%
		SELF-DISCLOSURE OF THE AUTHOR'S PUBLIC OR PERSONAL LIFE	5.3%
		INCLUSIVE PRONOUNS	71.1%
		REFERENCES TO THE READER	39.5%
		GIVING THE NON-RESEARCHER AN ACTIVE VOICE/DIRECT QUOTES FROM THE NON-RESEARCHER	23.7%
		FEATURES OF CONVERSATIONAL DISCOURSE	42.1%
		QUESTIONS	44.7%
		HUMOR	10.5%
		EXPLICIT SELF-REFERENCE	21.1%

Table 6.2 Qualitative results from the corpus analysis of professional science journalism texts

Theme	Strategy	Examples of strategy use	Other strategies also used in the example
Subject Matter	LEDE	New music download patterns appear to closely resemble epidemic curves for infectious disease, study finds (Geddes, 2021).	PRESENTING RESULTS/CONCLUSIONS, STANCE MARKERS, IMAGERY, ANNOUNCING THE NEW FINDING OR NEW CONTRIBUTION TO THE DISCIPLINE
		Boeing's Starliner astronaut taxi may not get off the ground this year after all (Wall, 2021).	STANCE MARKERS, PRESENTING RESULTS/CONCLUSIONS
	CONTEXTUALIZATION	We're used to grabbing sodas from vending machines, but what about steaks? Imagine if a vending machine could deliver fresh produce and other food the way that a conventional grocery store or bodega does. That way, you could quickly grab the one specific item you need without going indoors or interacting with other people during the pandemic (Hu, 2021).	INCLUSIVE PRONOUNS, QUESTIONS, EXAMPLES FROM DAILY LIFE, IMAGERY, REFERENCE TO THE READER, STANCE MARKERS
		For half a billion years or so, our ancestors sprouted tails. As fish, they used their tails to swim through the Cambrian seas. Much later, when they evolved into primates, their tails helped them stay balanced as they raced from branch to branch through Eocene jungles. But then, roughly 25 million years ago, the tails disappeared (Zimmer, 2021).	FEATURES OF CONVERSATIONAL DISCOURSE, INCLUSIVE PRONOUNS, IMAGERY, HYPERLINK, STANCE MARKERS
	ANNOUNCING THE NEW FINDING OR NEW CONTRIBUTION TO THE DISCIPLINE	Flowers can sense when a bumblebee is nearby and release a burst of perfume in order to attract more insects, scientists have found (Pinkstone, 2021).	PRESENTING RESULTS/CONCLUSIONS, IMAGERY, STANCE MARKERS
	NOVELTY	Previous research had identified two ancestor groups: hunter-gatherers who lived in Japan 15,000 years ago (and possibly much earlier) and farmers who migrated from East Asia starting around 900 B.C.E., reports Harry Baker for Live Science (Gershon, 2021).	STANCE MARKERS, ADDITIONAL SOURCES, HYPERLINKS
	DESCRIBING THE METHOD	To see what effect this shortened protein might have, Xia and his colleagues used the gene editor CRISPR to make mice with a shortened version of TBXT (Vogel, 2021).	STANCE MARKERS, LEXICAL MENTIONS
		To avoid damaging local ecosystems, conductors and circuit chips that degrade over time. After collecting and transmitting data about their landing zone, the fallen fliers disintegrate and melt into goo, which then washes away. This is better for the environment—and more convenient for researchers (Ogasa, 2021).	LEXICAL MENTIONS, IMAGERY
	PRESENTING RESULTS/ CONCLUSIONS	The new modeling shows a 3% chance that Lake Powell, which is located on the Colorado River from northern Arizona to southern Utah, could drop below the minimum level needed to allow the lake's Glen Canyon Dam to generate hydroelectricity next year. In 2023, the chance of a shutdown grows to 34%, according to the projection (Kann, 2021).	STANCE MARKERS, MENTION OF STATISTICS
		The researchers found that the builders of the ancient city did plenty of excavating—even quarrying bedrock for other construction sites in town—and that 65% of today's urban features are built on the same alignments as Teotihuacán. The team also found that 205 features from the ancient city have been destroyed by mining operations since 2015 (Schultz, 2021).	MENTION OF STATISTICS

(continued)

Table 6.2 (continued)

Theme	Strategy	Examples of strategy use	Other strategies also used in the example
Tailoring Information to the Reader	APPLIED IMPLICATIONS	Combined, these decisions will likely put a hard cap on the number of coal plants built and significantly limit the future export market for coal (Timmer, 2021).	N/A
		To avoid the negative impacts of technology, many choose to reassess their relationship with it and gradually reduce the time they spend using electronic devices. This awareness of living tech-dependent lifestyles and intention for better tech-life balances encourage people to undergo a "digital detox," the voluntary and intentional abstinence from technology use (Delgado, 2021).	HYPERLINKS, IMAGERY, EXPLANATIONS
	EXPLANATIONS	The other explanation points to stellar explosions called supernovas. When large stars run out of fuel and erupt in these violent supernovas, they can send nearby particles blasting away at near-light speed. These highly energetic particles, called cosmic rays, may then collide with other particles sprinkled through the gassy hinterland between stars, producing gamma-rays (Spector, 2021).	IMAGERY, STANCE MARKERS, NOVELTY, HYPERLINKS
	IMAGERY	Next to other airway-loving viruses, such as the ones that cause the flu and common colds, SARS-CoV-2 can be a bit of an oddball. It lopes almost indiscriminately throughout the body, invading a plethora of tissues; it winds up certain immune responses, while dialing others down, sparking bouts of inflammation that can afflict everything from brain to toe. COVID symptom lists that at first focused on the virus's ground zero—the respiratory tract—eventually ballooned to include nausea, vomiting, changes in mental status, and chest pain (Wu, 2021).	PRESENTING RESULTS/CONCLUSIONS, STANCE MARKERS, HYPERLINKS
	EXAMPLES FROM DAILY LIFE	What runs through your mind when you're deciding which toilet paper to buy? Sale price, roll size, pitiful single-ply or luxurious triple? Climate change might not make your list of considerations, but it should: … (Stuart-Ulin, 2021).	IMAGERY, REFERENCES TO THE READER, QUESTIONS, STANCE MARKERS, CONTEXTUALIZATION
	HYPERLINKS	COVID-19 vaccines are still largely keeping people out of the hospital in the United States, studies show (García de Jesús, 2021).	PRESENTING RESULTS/CONCLUSIONS
		The studies that have tackled the task of measuring real-world vaccine effectiveness against all symptomatic disease may not always count the same COVID-19 symptoms, experts told me, potentially inflating or deflating numbers (Wu, 2021).	DESCRIBING THE METHOD, EXPLICIT SELF-REFERENCE, IN-TEXT SPECIFICATION OF A SOURCE, STANCE MARKERS

			DESCRIBING THE METHOD
Visuals		 This is one of the visuals of Fifi the llama used in the text about immune therapy on llamas (Gill, 2021). Image credit: University of Reading.	
Credibility	ACADEMIC IMPLICATIONS	Automated tools likely still won't be used in the Decadal Surveys for some years to come. But if the survey committee does decide to integrate AI into its process, that will represent a new way for scientists to reach agreement on their own goals (Woodall, 2021).	N/A
	MENTIONING MORE RESEARCH IS NECESSARY/NEXT STEP IN RESEARCH	"[I]t's an exciting field of research, though still in its infancy," writes Kleinewietfeld, adding "more studies [are] needed to understand the complex interactions of nutrition, microbiome and immunity in the context of cancer. Thus, future studies will show if new findings could indeed lead to novel treatment options for patients" (Fessi, 2021).	STANCE MARKERS, LEXICAL MENTION, GIVING THE RESEARCHER AN ACTIVE VOICE, APPLIED IMPLICATION
	CONTRIBUTION TO RESEARCH	"We've been able to see these clouds for decades, but we never knew their true shape, depth, or thickness," Zucker, the lead author on a separate study that detailed the work set to be published in the Astrophysical Journal, said in the press release. "We also were unsure how far away the clouds were. Now we know where they lie with only 1 percent uncertainty, allowing us to discern this void between them" (Lea, 2021).	LEXICAL MENTION, HYPERLINK, IN-TEXT SPECIFICATION OF A SOURCE, STANCE MARKERS, MENTION OF STATISTICS
	MENTION OF STATISTICS	A previous, eye-popping ambition to grab 1% of emissions by 2025 is no longer on the cards (Removing carbon dioxide , 2021).	IMAGERY, PRESENTING RESULTS/CONCLUSIONS
		Overall, 80% of bird species studied showed changes in their counts in urban areas in the 2020 time frame—with most of them increasing on the order of 10% or 20% (Neuman, 2021).	PRESENTING RESULTS/CONCLUSIONS

(continued)

Table 6.2 (continued)

Theme	Strategy	Examples of strategy use	Other strategies also used in the example
	GIVING THE RESEARCHER AN ACTIVE VOICE/DIRECT QUOTES FROM THE RESEARCHER	They will also need to look at whether nepetalactone acts as a cat attractant as well as an insect repellent. "If you are walking around with this molecule on you, will there be no mosquitoes, but all the neighbourhood cats chasing you around? To be honest, I don't know," says Martin. "That's certainly something we're going to have to investigate" (Le Page, 2021).	MENTIONING MORE RESEARCH IS NECESSARY, REFERENCE TO THE READER, EXAMPLE FROM DAILY LIFE, QUESTIONS, LEXICAL MENTION, STANCE MARKERS. [I and we in this example refer to the researcher(s) and therefore are not coded as EXPLICIT SELF-REFERENCE or INCLUSIVE PRONOUNS]
	LEXICAL MENTION OF THE ORIGINAL RESEARCH	"So, you have this reserve," says John Wherry, an immunologist at the University of Pennsylvania Perelman School of Medicine in Philadelphia, who led the study. "Circulating antibodies may be declining, but your immune system is capable of jumping into action once again" (Dolgin, 2021).	GIVING THE RESEARCHER AN ACTIVE VOICE, REFERENCES TO THE READER, PRESENTING RESULTS/CONCLUSIONS
	ADDITIONAL SOURCES	Earlier this month, she was part of a team of researchers that examined breakthrough infections among health-care workers in San Diego. They noted that the drop in vaccine effectiveness from June to July was likely caused by waning immunity and the emergence of the Delta variant (McKeever, 2021).	HYPERLINK, STANCE MARKERS, IMAGERY
	LINK TO THE ACADEMIC PUBLICATION	The new laser is detailed in Nature Nanotechnology (Dumé, 2021).	HYPERLINK
	DIRECT QUOTE FROM THE ACADEMIC PUBLICATION	S-E Kim et al, Nat. Mater, 2021, https://doi.org/10.1038/s41563-021-01075-3 (Addison, 2021).	HYPERLINK
		Past studies have shown that rodents display strong neurophysiological responses to LSD. They tend to move less and start rapidly twitching their heads, which is a "behavioral signature of a hallucination-like state in rodents," the researchers noted (Johnson, 2021).	NOVELTY, STANCE MARKERS
	IN-TEXT SPECIFICATION OF A SOURCE	"Potassium is a moderately volatile element (mimicking the behavior of highly volatile elements), but it is not too volatile to get completely lost," wrote Kun Wang, a planetary scientist at Washington University and one of the authors of the study, in an email to TIME (Kluger, 2021).	GIVING THE RESEARCHER AN ACTIVE VOICE, EXPLANATION, STANCE MARKERS, LEXICAL MENTION
Stance	OPINION	In August, this evidence prompted the FDA to approve use of a booster dose for certain immunocompromised populations. "To the extent that a third shot gets them a little bit closer to what we see in healthy people after their second shot, I think that's worth doing," Bhattacharya says. "That's the easy one" (McKeever, 2021).	HYPERLINK, STANCE MARKERS, EXAMPLE FROM DAILY LIFE, MENTION OF STATISTICS, GIVING THE RESEARCHER AN ACTIVE VOICE, LEXICAL MENTION. [We and I refer to the researcher(s) and therefore are not coded as INCLUSIVE PRONOUNS or EXPLICIT SELF-REFERENCE]
	STANCE MARKERS	China in theory could develop an orbital nuclear weapon that could dodge America's mainly north-facing strategic radars (Axe, 2021).	IMAGERY, PRESENTING RESULTS/CONCLUSIONS
		Forests are, of course, good for the planet (Jones, 2021).	APPLIED IMPLICATION

Engagement	TITLES/SUBHEADINGS	We can now bio-engineer catnip instead of extracting it from plants (Le Page, 2021).	INCLUSIVE PRONOUNS, PRESENTING RESULTS/CONCLUSIONS
		Birds thrived where humans feared to tread during the pandemic, scientists say (Neuman, 2021).	PRESENTING RESULTS/CONCLUSIONS, STANCE MARKER
	REFERENCES TO POPULAR LORE AND BELIEFS, AND POPULAR CULTURE	While we don't expect to be delivering a full Star Trek holodeck experience in the near future, we're already boldly going in new directions to add additional functions to the system (Dahiya, 2021).	MENTIONING MORE RESEARCH IS NECESSARY, STANCE MARKERS. [This article is written by researchers; therefore 'we' refers to the researchers and is not coded as an INCLUSIVE PRONOUN]
	SELF-DISCLOSURE OF THE AUTHOR'S PUBLIC OR PERSONAL LIFE	For me, the pandemic anxiety that dominated much of 2020 is slow to fade, and the idea of getting COVID-19 still feels far worse than getting the flu, even if the symptoms were identical (Wu, 2021).	IMAGERY
	INCLUSIVE PRONOUNS	Pizza with leftover olive leaves. Bread doctored with rice waste. Banana peels turned into snacks. These are examples of a trend called upcycled food — and soon they are coming to our plates (Zaraska, 2021).	CONTEXTUALIZATION, EXAMPLES FROM DAILY LIFE, IMAGERY, HYPERLINK
	REFERENCES TO THE READER	Pop music is often described as catchy, but it seems you really can infect friends with your music taste (Geddes, 2021).	EXAMPLE FROM DAILY LIFE, CONTEXTUALIZATION, PRESENTING RESULTS/CONCLUSIONS, STANCE MARKERS
	GIVING THE NON-RESEARCHER AN ACTIVE VOICE/DIRECT QUOTES FROM THE NON-RESEARCHER	Young climate activists said feelings of anxiety over the climate were now widespread among today's youth. Mitzi Tan, 23, from the Philippines, said: "I grew up being afraid of drowning in my own bedroom. Society tells me that this anxiety is an irrational fear that needs to be overcome, one that meditation and healthy coping mechanisms will 'fix'. At its root, our climate anxiety comes from this deep-set feeling of betrayal because of government inaction. To truly address our growing climate anxiety, we need justice" (Harvey, 2021).	EXAMPLES FROM DAILY LIFE, IMAGERY, STANCE MARKERS, INCLUSIVE PRONOUNS, APPLIED IMPLICATION. [I is a reference to the active voice, not to the writer, and is therefore not coded as an EXPLICIT SELF-REFERENCE]
	FEATURES OF CONVERSATIONAL DISCOURSE	To understand that interaction, researchers had to, unfortunately, wait for a decent number of people to get sick-to observe the virus screwing with us in real time (Wu, 2021).	PRESENTING RESULTS/CONCLUSIONS, STANCE MARKERS, INCLUSIVE PRONOUNS, IMAGERY
		But it's not as easy as snapping your fingers and having everything work out (Koerth, 2021).	REFERENCE TO THE READER
	QUESTIONS	For example, why did large rocks tend to create tables while smaller rocks simply sunk down into the ice below? (Gammon, 2021). Disgusting? Not necessarily (Zaraska, 2021).	CONTEXTUALIZATION, STANCE MARKERS
			FEATURES OF CONVERSATIONAL DISCOURSE
	HUMOR	Dino-mite find! Bizarre armoured spike fossil belongs to a new species of dinosaur that lives in Africa 168 MILLION years ago and was 'unlike anything else in the animal kingdom' (Morrison, 2021).	TITLE, STANCE MARKERS, PRESENTING RESULTS/CONCLUSIONS, FEATURES OF CONVERSATIONAL DISCOURSE, GIVING THE RESEARCHER AN ACTIVE VOICE
	EXPLICIT SELF-REFERENCE	My colleagues and I working in the University of Glasgow's bendable electronics and sensing technologies research group have now developed a system of holograms of people using "acrohaptics", creating feelings of touch with jets of air (Dahiya, 2021).	LEXICAL MENTION, HYPERLINK, LINK TO THE ACADEMIC PUBLICATION, ANNOUNCING THE NEW FINDING OR NEW CONTRIBUTION TO THE DISCIPLINE, PRESENTING RESULTS/CONCLUSIONS, EXPLANATION
		As I've previously reported, grasslands store vast amounts of carbon — most of which is below ground — and provide homes for countless species (Jones, 2021).	HYPERLINK, STANCE MARKERS, PRESENTING RESULTS/CONCLUSIONS

6.4   Interpretation

After using the framework to analyze these 38 science journalism texts, the following insights can be gathered about the corpus. The average number of strategies that was used is 18.7 strategies per text, and the median is 20 strategies with the lowest score of 10 strategies and the highest of 27 strategies. This also means that no single text uses *all* available strategies. Several strategies appear in every single text in the corpus (100%): PRE-SENTING RESULTS/CONCLUSIONS, HYPERLINKS, VISUALS, STANCE MARKERS, and TITLES/SUBHEADINGS. Other strategies are used in almost all texts (75–99%): CONTEXTUALIZATION, ANNOUNCING THE NEW FINDING OR NEW CONTRIBUTION TO THE DISCIPLINE, and IMAGERY. In most texts (50–75%), the following strategies are used: LEDE, DESCRIBING THE METHOD, APPLIED IMPLICATIONS, EXPLANATIONS, GIVING THE RESEARCHER AN ACTIVE VOICE, LEXICAL MENTION OF THE ORIGINAL RESEARCH, ADDITIONAL SOURCES, and INCLUSIVE PRONOUNS. A few strategies are only used in some texts (25–50%): NOVELTY, EXAMPLES FROM DAILY LIFE, ACADEMIC IMPLICATIONS, MENTIONING MORE RESEARCH IS NECESSARY/NEXT STEP IN RESEARCH, CONTRIBUTION TO RESEARCH, MENTION OF STATISTICS, LINK TO THE ACADEMIC PUBLICATION, OPINION, REFERENCES TO THE READER, FEATURES OF CONVERSATIONAL DISCOURSE, and QUESTIONS. And lastly, some strategies are hardly used at all (0–25%): DIRECT QUOTE FROM THE ACADEMIC PUBLICATION, IN-TEXT SPECIFICATION OF A SOURCE, REFERENCES TO POPULAR LORE AND BELIEFS, AND POPULAR CULTURE, SELF-DISCLOSURE OF THE AUTHOR'S PUBLIC OR PERSONAL LIFE, GIVING THE NON-RESEARCHER AN ACTIVE VOICE, HUMOR, and EXPLICIT SELF-REFERENCE.

Looking at the use of strategies per theme, the themes Subject Matter (75.9%), Tailoring Information to the Reader (80.3%), and Stance (81.6%) are used most. Or, to explain it differently, the theme Subject Matter contains six strategies that could be used across 38 texts, which means that the corpus enables 228 'options' to use Subject Matter strategies, of which 173, or 75.9%, are in fact used. The aggregated scores for Credibility (42.8%) and Engagement (36.6%) are much lower. These aggregated results per theme of course do gloss over differences on the level of strategies. To give an example, for the theme Stance, STANCE MARKERS are used in 100% of the texts but OPINIONS are used in 'only' 63.2% of the texts—which then leads to an aggregated score of 81.6%. Especially in the themes Credibility and Engagement, there are big differences in the use of strategies, with some strategies being used across all texts (TITLES) while others are used hardly to not at all (SELF-DISCLOSURE OF THE AUTHOR'S PUBLIC OR PERSONAL LIFE).

Although we did not undertake this analysis to contrast professional journalism writing with student science journalism writing, anecdotally we can share that the use of strategies by professionals is much denser than is the case in the corpus that liberal education students wrote, which is discussed in Chap. 7. This might imply that professional writers have a better grasp on language use and are better equipped to utilize the options provided in the form of different strategies. The use of strategies also shows much more overlap, that is, multiple strategies being used at the same time. A visual representation of this overlap in strategies and richness in the use of strategies can be seen in the text that is discussed in Chap. 5, which was also part of the corpus discussed in this chapter.

When comparing the results of the analysis with Berezow's (2017) infographic on quality of science reporting, we see that those texts that are part of the quadrant from 'evidence-based reporting'/'almost always compelling science content' are more academic in register—especially the texts from the websites of Nature and Science—which is particularly visible through generally lower scores on the theme Engagement. Furthermore, many differences are visible between individual texts such as the length of texts, the number of strategies used within them, and the themes that are used the most.

A next step, which was not performed for this corpus, would be to analyze the specific use of each strategy. In other words, if a writer uses, for example, the strategy REFERENCE TO THE READER, what does that look like in the text? This step is performed for the corpus of student writing that is discussed in Chap. 7.

What You Have Learned in This Chapter

- This chapter presented the analysis of a corpus of science journalism texts written by professional science journalists.
- Science journalism articles from a whole range of science journalism outlets (based on a scale of being evidence-based and showing compelling science content) and from a spectrum of different topics/research fields were used.
- The analysis was quantitative and instrumental, meaning the focus was on how often each strategy is used in the text.
- Information is given about the overall number of strategies that were used in texts, on the percentage of use of each individual strategy, and about aggregated scores per theme.
- Aggregated scores for the themes Subject Matter, Tailoring Information to the Reader, and Stance are generally high, with the themes Credibility and Engagement scoring much lower.

REFERENCES

Addison, F. (2021, September 21). Camouflage crystals' colour change controlled with pressure. *Chemistry World*. https://www.chemistryworld.com/news/liquid-crystals-under-pressure-create-squid-like-camouflage/4014428.article

Axe, D. (2021, September 22). Yes, China could park nukes in orbit. America would have itself to blame. *Forbes*. https://www.forbes.com/sites/davidaxe/2021/09/22/yes-the-chinese-could-park-nukes-in-orbit-america-would-have-itself-to-blame/?sh=65844bde6381

Berezow, A. (2017, March 5). Infographic: The best and worst science news sites. *American Council on Science and Health*. https://www.acsh.org/news/2017/03/05/infographic-best-and-worst-science-news-sites-10948

Dahiya, R. (2021, September 17). We created holograms you can touch—You could soon shake a virtual colleague's hand. *The Conversation*. https://theconversation.com/we-created-holograms-you-can-touch-you-could-soon-shake-a-virtual-colleagues-hand-167478

Delgado, C. (2021, September 20). Technology overuse and the fear of "digital dementia": What you need to know. *Discovery*. https://www.discovermagazine.com/health/technology-overuse-and-the-fear-of-digital-dementia-what-you-need-to-know

Dolgin, E. (2021, September 17). COVID vaccine immunity is waning—How much does that matter? *Nature*. https://www.nature.com/articles/d41586-021-02532-4

Dumé, I. (2021, September 22). Moiré superlattice makes magic-angle laser. *Physics World*. https://physicsworld.com/a/moire-superlattice-makes-magic-angle-laser/

Fessi, S. (2021, September 20). Salty diet helps gut bugs fight cancer in mice: Study. *The Scientist*. https://www.the-scientist.com/news-opinion/a-salty-diet-helps-gut-bugs-fight-cancer-in-mice-study-69197

Gammon, K. (2021, September 15). Stone cold: How rocks become glacial tables. *Inside Science*. https://www.insidescience.org/news/stone-cold-how-rocks-become-glacial-tables

Garcia de Jesús, E. (2021, September 21). Why only some people may get COVID-19 booster shots at first. *Science News*. https://www.sciencenews.org/article/covid-coronavirus-who-gets-booster-shots-vaccines-pfizer-fda

Geddes, L. (2021, September 22). Mathematicians discover music really can be infectious—Like a virus. *The Guardian*. https://www.theguardian.com/science/2021/sep/22/mathematicians-discover-music-really-can-be-infectious-like-a-virus

Gershon, L. (2021, September 21). DNA analysis rewrites ancient history of Japan. *Smithsonian*. https://www.smithsonianmag.com/smart-news/japanese-ancestors-came-from-three-ancient-groups-180978725/

Gill, V. (2021, September 22). Covid: Immune therapy from llamas shows promise. *BBC*.. https://www.bbc.com/news/science-environment-58628689

Harvey, F. (2021, September 15). Anxious about climate, 4 in 10 young people are wary of having kids. *Mother Jones*. https://www.motherjones.com/environment/2021/09/global-survey-climate-change-anxiety-young-people-children-kids/

Hu, C. (2021, September 22). Meat vending machines are just the latest way the pandemic has reinvented eating. *Popular Science*. https://www.popsci.com/technology/restaurants-vending-machines-pandemic/

Johnson, S. (2021, September 22). LSD hallucinations are due to abnormal brain communication. *Big Think*. https://bigthink.com/mind-brain/lsd-hallucinations/

Jones, B. (2021, September 22). The false promise of massive tree-planting campaigns. *Vox*. https://www.vox.com/down-to-earth/22679378/tree-planting-forest-restoration-climate-solutions

Kann, D. (2021, September 23). There's a 1-in-3 chance Lake Powell won't be able to generate hydropower in 2023 due to drought conditions, new study says. *CNN*. https://edition.cnn.com/2021/09/23/weather/lake-powell-power-generation-outlook/index.html

Kluger, J. (2021, September 21). Mars was always destined to die. *Time*. https://time.com/6100276/mars-water-loss/

Koerth, M. (2021, September 8). Vaccines mandates work, but they're messy. *Five Thirty Eight*. https://fivethirtyeight.com/features/vaccines-mandates-work-but-theyre-messy-business/

Le Page, M. (2021, September 21). We can now bioengineer catnip instead of extracting it from plants. *New Scientist*. https://www.newscientist.com/article/2290867-we-can-now-bioengineer-catnip-instead-of-extracting-it-from-plants/

Lea, R. (2021, September 22). Astronomers discover mysterious 500-light-year-wide void in space: 'Absolutely shocking'. *Newsweek*. https://www.newsweek.com/astronomers-discover-mysterious-500-light-year-wide-void-cavity-space-1631576

McKeever, A. (2021, September 17). Why you may not need a COVID-19 booster yet after all. *National Geographic*. https://www.nationalgeographic.com/science/article/why-you may-not-need-a-covid-19-booster-after-all

Morrison, R. (2021, September 23). Dino-mite find! Bizarre armoured spike fossil belongs to a new specifies of dinosaur that lives in Africa 168 MILLION years ago and was 'unlike anything else in the animal kingdom'. *Daily Mail*. https://www.dailymail.co.uk/sciencetech/article-10018043/Bizarre-armoured-spike-fossil-belongs-dinosaur-lived-Africa-168-MILLION-years-ago.html

Neuman, S. (2021, September 22). Birds thrived where humans feared to thread during the pandemic, scientist say. *NPR*. https://www.npr.org/sections/coronavirus-live-updates/2021/09/22/1039593706/birds-thrived-in-urban-settings-during-pandemic-lockdowns

Ogasa, N. (2021, September 22). Winged microchips glide like tree seeds. *Scientific American*. https://www.scientificamerican.com/article/winged-microchips-glide-like-tree-seeds/

Pinkstone, J. (2021, September 22). Flowers can feel when bees are near and emit more scent, scientists find. *The Telegraph*. https://www.telegraph.co.uk/science/2021/09/22/flowers-can-feel-bees-near-emit-scent-scientists-find/

Removing carbon dioxide from the air. The world's biggest carbon-removal plant switches on. (2021, September 18). The Economist. https://www.economist.com/science-and-technology/2021/09/18/the-worlds-biggest-carbon-removal-plant-switches-on

Schultz, I. (2021, September 21). Archeologists see ancient Teotihuacán with aerial mapping tech. *Gizmodo*. https://gizmodo.com/archaeologists-see-ancient-teotihuacan-with-aerial-mapp-1847716948

Specktor, B. (2021, September 21). Scientists finally have an explanation for the most energetic explosions in the universe. *Live. Science*. https://www.livescience.com/empty-sky-gamma-ray-burst-supernova-emissions

Stuart-Ulin, C. R. (2021, September 18). The toilet paper companies destroying Canada. *Slate*. https://slate.com/technology/2021/09/toilet-paper-charmin-trees-canada-boreal.html

Timmer, J. (2021, September 22). China to stop building coal plants in developing nations. *Ars Technica*. https://arstechnica.com/science/2021/09/china-to-stop-building-coal-plants-in-developing-nations

Vogel, G. (2021, September 21). 'Jumping gene' may have erased tails in humans and other apes – and boosted our risk of birth defects. *Science*. https://www.science.org/content/article/jumping-gene-may-have-erased-tails-humans-and-other-apes-and-boosted-our-risk-birth-defects

Wall, M. (2021, September 22). Boeing's next Starliner test launch for NASA may slip to 2022. *Space*. https://www.space.com/boeing-starliner-oft-2-test-launch-may-slip-2022

Woodall, T. (2021, September 21). This AI could predict 10 years of scientific priorities—If we let it. *MIT Technology Review*. https://www.technologyreview.com/2021/09/20/1035890/ai-predict-astro2020-decadal-survey/

Wu, K. J. (2021, September 21). 'Post-Vax COVID' is a new disease. *The Atlantic*. https://www.theatlantic.com/science/archive/2021/09/post-vaccination-covid/620140/

Zaraska, M. (2021, September 18). Upcycling food waste onto our plates is a new effort. But will consumers find it appetizing? *Washington Post*. https://www.washingtonpost.com/science/upcycling-food-waste/2021/09/17/90fd81b2-0045-11ec-85f2-b871803f65e4_story.html

Zimmer, C. (2021, September 21). How humans lost their tails. *The New York Times*. https://www.nytimes.com/2021/09/21/science/how-humans-lost-their-tails.html

Corpus Example: Using an Analytical Framework to Characterize First-Year Undergraduate Newspaper Article Writing

Abstract This chapter details the analysis of a corpus of 140 newspaper articles based on a single academic source text written by first-year undergraduate liberal education students. The analysis is both quantitative and representational, and qualitative and inductive. It is quantitative in the sense that data were produced about the number of times each strategy was employed. In a second, qualitative step, the way the student writers employ each strategy is determined and categorized. In the original analysis, all 34 strategies in the analytical framework were analyzed, but to keep this chapter readable, we only present the results for one strategy out of each theme: DESCRIBING THE METHOD from Subject Matter, APPLIED IMPLICATIONS from Tailoring Information to the Reader, MENTION OF STATISTICS from Credibility, STANCE MARKERS from Stance, and REFERENCES TO THE READER from Engagement. To conclude this chapter, we discuss the general insights that can be drawn from this analysis and consider the use of the analytical framework in didactic settings.

Keywords Science journalism • Student writing • Text analysis • Educational settings

7.1 Introduction

This chapter presents the results of the analysis of a corpus of newspaper articles written by undergraduate students. This analysis is both quantitative and representational as well as qualitative and inductive: It shows *how*

© The Author(s) 2023
F. M. Sterk, M. M. van Goch, *Re-presenting Research*,
https://doi.org/10.1007/978-3-031-28174-7_7

often each strategy was used, and *in what way* each strategy was used. The goal of the chapter is to give insight into what kinds of results can be expected when using the framework to analyze a corpus of texts in-depth. Furthermore, because this corpus was written by first-year students, this chapter shows that the framework can also be used in didactic or educational settings. For readability, we have limited the presentation of results to the analysis of five strategies.

7.2   Corpus Construction

This corpus consists of 140 newspaper articles written by first-year undergraduate liberal education students. These students were enrolled in the program Liberal Arts and Sciences at Utrecht University, The Netherlands; students pick their own specialization and receive training in interdisciplinary research skills to become so-called disciplined interdisciplinarians. Students were asked to write a newspaper article in the first month of their first year of training, at a point in the program when they have not declared a disciplinary specialization. Although the core curriculum of Liberal Arts and Sciences places a strong focus on *academic* writing, no specific training in popularization writing is offered until the final year of the program. As the program offers education in Dutch, all texts were written in Dutch and examples in this chapter have been translated into English for readability.

Whereas the corpus in Chap. 6 is based on a diverse range of source texts, topics, and disciplines, this corpus is based on a single source text: "#Sleepyteens: Social media use in adolescence is associated with poor sleep quality, anxiety, depression and low self-esteem" (Woods & Scott, 2016). This academic text details the effects of social media use, especially during the night, on the lives of teens. Students were asked to read this academic text before class. In class, students were given one hour to write a newspaper article of 400 words about the academic paper. The newspaper article had to be suitable to publish in the science section of a quality Dutch newspaper. The target audience of the text consisted of a general audience that was interested in science, but did not necessarily receive higher education training. Students were told that the goal of their text was to retain the readers' interest and make sure that the presented insights were understandable. Part of the resulting corpus was used in the construction phase in the development of the analytical framework (see Chap. 4).

7.3 Analysis

The analysis was conducted in two steps. In the first step, each individual text was analyzed for the occurrence of strategies, using the analytical framework. This led to a type of analysis and outcome that was also shown in Chap. 5. For each text, all occurrences were coded in NVivo. This led to quantitative and representational data; in other words, for each strategy, a percentage could be given to denote how many texts made use of it. This data is visualized in tables (see Table 7.1 through 7.5) in which the number of texts that used a strategy is shown as well as the number of references, or in other words how many times that strategy was used in how many texts. Because a single strategy can be employed multiple times in one text, the number of references can be higher than the total number of texts.

The second step was qualitative and inductive; per strategy, all coded text was gathered in NVivo and then coded again in subforms (or subcodes). In other words, the *way* each strategy was used in this specific corpus was characterized. For example, what does the use of the strategy HUMOR look like? How do students use STATISTICS? By thematizing and describing how students employed the strategies, this step provided us with insights into the way first-year liberal education students write popularization texts. The insights the analysis produced about student popularization writing skills and consequences for setting up an educational program in popularization writing are detailed in Sterk et al. (2022). Here, on the other hand, we will focus on the textual side of the analysis and show the different ways in which strategies were employed within this corpus. With this information, Tables 7.1, 7.2, 7.3, 7.4, and 7.5 were expanded with the number of texts and references for each specific use of a strategy. For reasons of readability and succinctness, in this chapter we only reproduce the results for five of the 34 strategies—one from each theme: DESCRIBING THE METHOD, APPLIED IMPLICATIONS, MENTION OF STATISTICS, STANCE MARKERS, and REFERENCES TO THE READER. The other strategies were analyzed in the same manner but are not reported on in this chapter. If context is added for clarity in the example text, the strategy under discussion is underlined.

7.4   ANALYSIS EXAMPLE 1: DESCRIBING THE METHOD

The results of the analysis for the Subject Matter strategy DESCRIBING THE METHOD are shown in Table 7.1. Seventy-one percent of students used DESCRIBING THE METHOD to add information about the methodology of the academic text in their newspaper article. For this strategy, five subcodes were found: participants, research question/goal/hypothesis, measured constructs, materials, and mentions of the original research.

The level of specificity varies, with some students re-presenting the methodological framework word for word and others writing a summarized statement. Students pick and mix information from the academic text; in most texts, multiple subcodes are represented. The next example combines statements about the participants, the research question, and the materials:

(1) The research into the influence of social media among teens [research question] was conducted in Scotland. 467 Scottish pupils between 11 and 17 years old took part [participants]. In class and online, questionnaires were administered that teens needed to fill in [materials].

The subcode that appears most often is that of participants, who are mentioned in varying degrees of specificity. The academic article explains that

(2) Participants were 467 Scottish secondary school pupils aged 11–17 years. (Woods & Scott, 2016, p. 43)

Most student texts mention that 467 Scottish teens took part in the study, some add they were aged 11–17, others merely mention that 'about 500 participants' took part:

Table 7.1   Results for DESCRIBING THE METHOD

	Number of texts	Number of references
DESCRIBING THE METHOD	100	153
Participants	76	86
Research question/goal/hypothesis	68	85
Measured constructs	37	39
Materials	32	43
Mentions of the original research	16	19

(3) The research was conducted with students from secondary school in Scotland, they were in the age group of 11 to 17 years old.

The research question and hypothesis appear in student texts in multiple different forms. In the academic text, the research question is posed in the following way:

(4) The present study makes a novel contribution to the literature by examining how overall vs. night-time specific social media use and emotional investment in social media relate to sleep quality, anxiety, depression and self-esteem in adolescents. (Woods & Scott, 2016, p. 41)

Some student writers stick close to the academic text and mention night-time versus overall social media use or sleep quality, anxiety, depression, and self-esteem when talking about the research question. Others represent these statements in a more abstract manner and mention that research is being done into the effects of social media. Generally, all texts that mention the research question include a mention about social media. In the following example, the same constructs are mentioned as in the academic text, but the research question is reformulated to be easier to understand:

(5) A connection is sought between the use of social media and the quality of sleep, and with depression, anxiety, and self-esteem. The focus is specifically on the influence of the use of social media right before going to bed.

In the academic text, the measured constructs mentioned are poor sleep quality, anxiety, depression, self-esteem, emotional investment in social media, overall social media use, and night-time-specific social media use (Woods & Scott, 2016). Each construct has its own subheading, making it easily discoverable by the reader, and a short paragraph of text that explains the materials used to measure these constructs. In the student corpus, some writers mention all constructs, yet more often, only a selection of constructs is disclosed:

(6) These students were asked about their sleep quality, their mental health (with regards to depression and anxiety), their self-esteem,

how much social media they use and how much of this is at night, and how big their emotional investment in social media is.

In the academic text, the references made to materials center around the use of a questionnaire:

(7) Pupils in 1st to 4th year (aged 11–15) completed questionnaires in class, either in pencil-and-paper form or online, hosted by qualtrics.com. … Pupils in 5th and 6th year (aged 15–17) completed the online questionnaire hosted by qualtrics.com outside of class, via a link circulated by the school. (Woods & Scott, 2016, p. 43)

In the student writing, some texts only mention the questionnaire; others add that it was conducted partially online and partially in class, or mention the type of scale that was used to measure outcomes, or that consent was obtained:

(8) In class as well as online, questionnaires were provided that teens needed to fill in.

Comments about the methodology are a logical place to not only show how but also by whom the research was conducted. Mentions include the authors of the paper, the location of the research, and the title of the paper. These types of mentions were also coded in another strategy: LEXICAL MENTION OF THE ORIGINAL RESEARCH.

(9) The research '#Sleepyteens: Social media use in adolescence is associated with poor sleep quality, anxiety, depression and low self-esteem' investigates the influences of social media of teens on the sleep quality and psychological well-being of teens.

7.5   ANALYSIS EXAMPLE 2: APPLIED IMPLICATIONS

The results of the analysis for the strategy APPLIED IMPLICATIONS, part of the theme Tailoring Information to the Reader, are shown in Table 7.2. Forty-eight percent of students mentioned APPLIED IMPLICATIONS of the insights of the academic text in their newspaper article. Two types of APPLIED IMPLICATIONS are used: implications and calls to action. An implication

Table 7.2 Results for APPLIED IMPLICATIONS

	Number of texts	Number of references
APPLIED IMPLICATIONS	67	82
Call to action—analyzed by content	56	62
Conscious choices social media	23	23
No mobile phone	20	21
Conscious choices social media + no mobile phone	8	8
Not further specified	10	10
Call to action—analyzed by referent	56	64
For the reader	20	24
For 'us'	13	14
For teens	11	11
No referent	8	8
For parents	3	4
Using self-reference	3	3
Implications	18	21
For teens	11	11
For mental health	4	4
For society	4	4
For parents	2	2

represents the further-reaching consequences of the research topic. A call to action nudges the reader into taking a certain action needed because of those implications. In most cases, when a call to action is added, the implication on which the call to action is based is left implicit in the student writing.

The first type of APPLIED IMPLICATION is implications. Implications are mentioned on the level of teens, mental health, society, and parents, though never on the level of research because that type of implication is part of a different strategy: ACADEMIC IMPLICATIONS. In the student corpus, mentioned implications are about teens, mental health, society, or parents. Often, information that is presented as a result in the academic source text is instead represented as an implication in student writing. On the one hand, this shows that participants stick close to the source text. On the other hand, this poses a factual misrepresentation of the source text because it did not specify the information to be an implication.

For teens, the focus is on the potential implications of social media and the use of smartphones at night, for example on health. For this subcode specifically, what is presented as an implication in the student writing is in fact part of the results in the academic source text.

(10) It is the future, and sometimes the unfortunate future, that this generation of teens will grow up with constant connectivity and experience insomnia as being unchangeable.

Implications for mental health similarly feature claims that originate as results from the research but are re-presented as implications. In the following example, the mental and emotional low point that is mentioned shows a connection to one of the consequences of social media use that is discussed in the academic text:

(11) Not only has our physical condition deteriorated because of intensive and late-night social media use, it also brings us to a mental and emotional low point.

Implications for society do present implications that are further reaching than the information already presented in the paper. Here, claims are made about the implications for society at large:

(12) Right now, it is important to do research as a society to learn how to deal with these consequences. It is important that technology keeps striving forward, but this should not be at the cost of the well-being of humans.

Implications for parents show the action that parents can undertake because of the new information in the academic paper. The difference with a call to action that is directed at parents (see below) is that a call to action explicitly and directly speaks to parents to spur them on to take action, whereas implications do not directly address the parents:

(13) Because of this research they [parents] can warn their brood with more conviction about the consequences that social media have.

The second type of APPLIED IMPLICATION is a call to action. Calls to action can be classified either by content (that is, the action they refer to) or by referent (that is, the person that needs to take action). In terms of content, broadly seen, a call to action can contain either one or two claims that are both closely connected to the main claim. The first claim is that conscious choices need to be made about the amount of time spent on social media. This call to action connects to claims in the results of the academic text that state teens use a lot of social media, and that this might cause unwanted consequences. In these calls to action, the reader in general or teens specifically are addressed and asked to consider the amount of time they spend on social media:

(14) Use it [social media] the way you like it, but when it has an influence on wellbeing, it's sensible to 'disconnect' for a while and enjoy your beautiful experiences without the nasty side-effects of social media.

The second claim made in calls to action is that readers should not use a mobile phone late at night or should not take their phone to bed. This call to action connects more specifically to the claim in the academic paper that 86% of teens take their smartphone to bed with them:

(15) So just leave your phone downstairs when you go to bed, and it really doesn't hurt you to send a little less texts every now and again.

The third type of call to action combines the two claims about making conscious choices about social media and leaving mobile phones out of the bedroom:

(16) That is why it is wise to take a step back from social media. The most important step is to leave your mobile phone and laptop downstairs. That way you're not tempted to use social media at night. It's also important that you're not too invested in social media.

The fourth subcode contains calls to action that are not further specified. This subcode contains calls to action in the realm of social media and phone use that might propose a change on a higher level than the individual:

(17) We should think of an alternative to deal with this [social media].
A way that doesn't negatively impact mental health and the
quality of life.

Students use six different referents in their calls to action: the reader,
us, teens, calls to action where the referent is not specifically addressed,
parents, and self-reference. Referents are mostly teens or people connected
to them, which shows a clear connection to the topic of the research.

A call to action is most often addressed to the reader in general, usually
by using 'you,' although imperative verbs sometimes lack a referent. These
are the most direct types of call to action. Although some of these texts are
clearly written with parents or teenagers in mind as the target audiences,
in the 20 texts where 'you' is used in the call to action, there is no clear
target audience distinguishable, which makes it difficult to specifically pin
down what type of reader 'you' refers to:

(18) So, next time you pick up your phone, or watch another Facebook
video, think twice, and put it away.

A call to action can also use 'we,' 'our,' and 'us' to make an inclusive
group of the author of the text and its readers. This type of referent is used
when the writer wants everyone to commit to the call to action.

(19) Maybe it's time for us to find a replacement for all the cat videos
and 'Facebook posts' and to start reading books again.

Calls to action can also be used to make a claim about the behavior of
teens that needs to change. In these texts, it is clear that teens themselves
are not the target audiences, but that the text is instead written about
teens and aimed at parents or maybe even the general audience.

(20) Teens should not be on social media just before going to bed and
especially not in the middle of the night when they wake up.

Calls to action can also be presented without a specific referent. In this
subcode, more general calls to action can be found that focus on a level
that overarches the individual:

(21) In the current society more awareness about the negative consequences of social media is necessary.

Similarly, parents can be the referent in the call to action. Although it is the parents that are spurred into taking action, the action still refers to the use of mobile phones and social media in teens:

(22) Make sure that <u>your teen</u> limits the use of social media and reduces it to zero in the evenings. Because of this, teens will lead a healthier life with less social and mental problems.

In some cases, the writer uses self-reference in a call to action. In these cases, the writer shows the reader the desired behavior through self-reference (the underlined words in this example are also coded for EXPLICIT SELF-REFERENCE):

(23) To conclude <u>I</u> would like to offer a resort for those people for whom it's not too late yet. All these problems originate in attaching too much value in social media. <u>My</u> tip is to be more aware of the time you spend online. Otherwise, you might be the next social junkie.

7.6   ANALYSIS EXAMPLE 3: MENTION OF STATISTICS

The results of the analysis for the Credibility strategy MENTION OF STATISTICS are shown in Table 7.3. Thirty-nine percent of students mentioned STATISTICS in their newspaper article. There are 11 types of STATISTICS that are used, captured under two main types of statements.

MENTIONS OF STATISTICS appear mostly in percentages, though sometimes they are written out in words. MENTIONS OF STATISTICS are exclusively used in two types of statements about the topic of the academic source text: either about issues teens face because of phone/social media use or about the behavior of teens regarding that phone/social media use. Although these claims are presented separately, one STATISTIC is often used in both claims; 97% of teens use social media, which is part of the claim about behavior of teens but also often added to claims about issues teens face. Most MENTIONS OF STATISTICS (96 out of 106) are directly lifted from the academic text; in only 10 cases did STATISTICS appear that were added from other sources.

In claims about issues teens have, five claims can be distinguished that are all mentioned in the academic text. Three of these STATISTICS are copied

Table 7.3 Results for MENTION OF STATISTICS

	Number of texts	Number of references
MENTION OF STATISTICS	54	109
Issues teens have	29	56
21% depressed	13	14
35% bad sleepers	13	13
47% anxiety	12	12
Not further specified	8	9
25% wake up from phone	5	5
37% loss of sleep through social media	3	3
Behavior of teens	41	50
90% use social media night and day	18	18
97% use social media	13	16
86% sleep with phone	11	11
54% of day on social media	6	6
Not further specified	1	1

out of the first paragraph of the results section in the academic article: 47% of teens suffer from anxiety, 35% of teens are classified as bad sleepers, and 21% are depressed. This information appears in the academic text in the following way:

> (24) Mean scores and standard deviations for each measure are presented in Table 1. 97% of participants indicated that they used social media. 35% of participants were classed as poor sleepers, with a PSQI score greater than 5... . PSQI scores were positively skewed, so were transformed—by taking log 10(score + 1)—to meet normality assumptions for all further analysis. 47% of participants were classed as anxious and 21% as depressed, according to the HADS cut-off score of 8 or above... . (Woods & Scott, 2016, p. 44)

In the popularized texts, the student writers ignore the explanation of the statistical analysis and reduce the content to simple STATISTICS, which are used only in the presentation of the results of the study—no other claims or consequences are drawn from the STATISTICS:

(25) 467 Scottish pupils were examined in this study. 97% of the pupils said they use social media. 35% report poor sleep quality. 47% of the pupils were classified as anxious and 21% as depressed.

Two other MENTIONS OF STATISTICS in claims about issues that teens face are not results from the research itself, but STATISTICS that are mentioned in the introduction of the academic text and originate in other academic sources: 37% of teens experience loss of sleep through social media and 25% of teens wake up because of their phone. The 37% STATISTIC is presented in the academic text as follows:

(26) Concerning social media in particular, Espinoza ... surveyed 268 young adolescents and found that 37% reported losing sleep due to the use of social networking sites. (Woods & Scott, 2016, p. 42)

This STATISTIC is mostly used as part of the strategy PRESENTING RESULTS/CONCLUSIONS, but can also be found in other parts of the text, for example as part of the strategy NOVELTY:

(27) A major change in the sleep pattern of youngsters has occurred with the introduction of social media. Teens sleep less because they spend more time on their phones when they are already in their beds. 37% indicate a lack of sleep with chronic fatigue as a consequence.

The 25% STATISTIC is presented in the following way in the academic text:

(28) A quarter of adolescents' report sleep interruptions from incoming text messages ... and social media alerts are likely to cause similar sleep disturbances. (Woods & Scott, 2016, p. 42)

This STATISTIC is presented in an almost identical fashion in student writing:

(29) A quarter of all teens indicate they sometimes wake up from notifications from their phone.

Some MENTIONS OF STATISTICS appear that the student writers added from other sources (10 out of 106 cases). In some texts, these other sources are mentioned, but not in all. In seven cases, these STATISTICS are presented as part of the strategy PRESENTING RESULTS/CONCLUSIONS, in two cases as part of the strategy NOVELTY and once in a CONTEXTUALIZATION. The STATISTIC used in the CONTEXTUALIZATION is:

(30) Almost <u>one in three</u> Dutch people indicate that they had trouble sleeping this year.

Here, the STATISTIC is not only used to introduce the topic but also to connect it to (Dutch) readers.

The second way in which MENTIONS OF STATISTICS are used is to make claims about behavior of teens. Here, four claims are made that are all part of the academic text: 97% use social media, 90% use social media at night and in the daytime, 86% sleep with their phone, and 54% of teens' days are spent on social media.

The fact that 97% of teens use social media is always included as a research result and re-presented together with the three STATISTICS about problems that teens experience, as already described. In fact, the 97% STATISTIC can already be seen in example (23). It is also presented on its own, without the STATISTICS about issues teens face:

(31) Out of all the participants in the research, <u>97%</u> indicated that they use social media.

The STATISTIC that 90% of teens use social media at night is mentioned in the opening sentence of the academic paper:

(32) Social media sites—such as Facebook and Twitter—have rapidly become a central part of young people's lives, with over 90% now using social media, day and night... . (Woods & Scott, 2016, p. 41)

In the corpus, the STATISTIC is represented as part of other strategies: seven times as part of NOVELTY, six times as part of CONTEXTUALIZATION, and five ties as part of PRESENTING RESULTS/CONCLUSIONS. The information is used in different ways; some texts mention that social media is used at night and in the daytime; others re-use the examples that are given of social media platforms and merely mention that 90% of teens use social media. The next example shows the STATISTIC being used as part of NOVELTY to mention that even though the use of social media is a pressing problem, not a lot of research is conducted into it:

(33) 90% of teens use social media, like Facebook and Twitter. So, it's no wonder that teens experience issues and the connection to social media is quickly made. But until recently, no research was done, and the connection was not proven yet.

The STATISTIC that 86% of teens sleep with their phone in bed is mentioned in the introduction of the academic text, as is the STATISTIC that teens spend 54% of their day on social media:

(34) Firstly, incoming alerts during the night have the potential to disturb sleep, as 86% of adolescents sleep with their phone in the bedroom—often under their pillow or in their hand... . (Woods & Scott, 2016, p. 42)

(35) Previous findings on Internet use in general are certainly relevant when considering social media use specifically, as young people spend 54% of their time online using social media... . (Woods & Scott, 2016, p. 42)

Both MENTIONS OF STATISTICS are mainly used as part of the strategy PRESENTING RESULTS/CONCLUSIONS, even though the STATISTICS are part of earlier research and not an actual result of Woods and Scott's (2016) research. The 86% STATISTIC is used as part of the strategy NOVELTY twice, and the 54% STATISTIC is used as part of the strategy NOVELTY twice and CONTEXTUALIZATION once. As part of PRESENTING RESULTS/CONCLUSIONS, it is used in the following way:

(36) What's more, the use of social media leads to the expectation that someone is alert all the time and so teens spend 54% of their time on social media.

In the next example, the 86% STATISTIC is used as part of the strategy NOVELTY. Here, the STATISTIC is deployed to introduce the topic, that is, it is the first piece of information that is mentioned about it:

(37) Checking your Instagram in the evening? <u>86%</u> of adolescents sleep with their phone in their room, under their pillow or in their hand. In <u>a quarter</u> of teens their sleep is disrupted by incoming messages.

When the STATISTIC is used as part of PRESENTING RESULTS/CONCLUSIONS, it is used as a supporting argument to underpin the findings from the study:

(38) This [study] showed that the use of social media has a negative influence on the confidence of adolescents and causes anxiety and depression. Therefore, the quality of sleep and the hours of sleep deteriorate, especially with night-time use of social media. The latter is more common than many people may think; <u>86%</u> of adolescents sleep with their phone in the bedroom, of which a considerable part sleeps with the phone under their pillow or in their hand.

Note that in general, when MENTIONS OF STATISTICS are used that are part of earlier research and not an actual result of Woods and Scott's (2016) research, in almost all cases they are incorrectly re-presented because students treat them as if they are results from the research even though they are supporting information from previous sources.

7.7   ANALYSIS EXAMPLE 4: STANCE MARKERS

The results of the analysis for the Stance strategy STANCE MARKERS are shown in Table 7.4. Seventy-one percent of student writers use STANCE MARKERS in their newspaper article. STANCE MARKERS are used throughout the corpus to comment on the research, its findings, or implications. They can be used to align with expectations, but also to go against expectations, or to signal insecurity. They are more often used to contextualize information than they are used to exert a value judgment or feelings. There are 14 types of STANCE MARKERS, about value, order of magnitude, aligning with expectations, insecurity, knowing for sure, deviating from expectation, awareness of information, contrast, denominator of time, explanation, commitment to a statement, signaling a reason, and giving focus.

Table 7.4 Results for STANCE MARKERS

	Number of texts	Number of references
STANCE MARKERS	100	200
Value	43	53
Order of magnitude	24	27
Align with expectations	21	23
Insecurity	14	14
Know for sure	14	14
Deviate from expectation	13	15
Awareness of information	12	14
Contrast	11	13
Denominator of time	6	9
Explanation	5	5
Commitment to a statement	6	7
Signal a reason	4	6
Giving focus	4	4

STANCE MARKERS such as 'interesting,' 'shocking,' 'important,' 'crucial,' 'special,' 'essential,' and 'ridiculous' can denote value.

(39) These are <u>serious</u> numbers.
(40) We all spend a <u>ridiculous</u> amount of time check our Instagram feed.

The order of magnitude is made clear through markers like 'strong,' 'big,' 'massive,' 'whopping,' 'a lot,' 'gigantic,' and 'huge.'

(41) In a <u>whopping</u> 35% of this group...
(42) This has a <u>gigantic</u> negative influence.

Statements can align with expectations, and while the expectations themselves are often not explicitly stated, the alignment of the results found in the study to those expectations is commented on, by using 'as expected,' 'it's not a surprise,' and 'indeed.' Alternatively, comments can denote a deviation from expectations, signaled by 'actually' and 'what's more.'

(43) <u>As expected</u>, a connection was found between the use of social media and the wellbeing of these students.

(44) All these small research projects have, <u>not totally unsurprisingly</u>, given the insight that...

(45) It can be concluded that social media <u>actually</u> does have an influence on sleep deprivation.

(46) <u>What's more</u>, social media use can also lead to depression, more stress, and less self-confidence.

Markers of insecurity show that something might be the case, but that the writer is not sure about it, using the markers 'would,' 'might,' 'maybe,' and 'possibly.'

(47) That this doesn't do much harm, is <u>maybe</u> a thought that you recognize yourself in.

(48) This happens more often than people <u>might</u> think.

Alternatively, if writers know something for sure, they can signal this by using 'absolutely,' 'clearly,' and 'certainly.'

(49) This would <u>certainly</u> have a positive effect on the sleep routines of students.

(50) The research <u>clearly</u> shows that the value that adolescents attach to social media contributes more to anxiety, depression, and a low self-esteem than the amount of exposure to it does.

The underlying assumption that information is in fact already known or accepted is denoted using 'of course.' These kinds of STANCE MARKERS explicate an underlying assumption that information that is being shared in the text is in fact already known by readers or accepted to be true:

(51) <u>Of course</u>, excessive social media use is a key factor in increasing sleep deprivation and depression, amongst others, in adolescents.

Contrast can be signaled by using the markers 'but' and 'however.'

(52) What was new in the research, <u>however</u>, was the specific focus on night-time social media use and emotional investment of teens in social media.

(53) That the use of a phone makes it difficult for you to get some sleep is not a secret, <u>but</u> that is not the only problem.

Denominators of time are used to show something is already happening or information is already known, using 'for some time' or 'quickly.'

(54) That this can have negative effects, had also been known <u>for some time</u>.
(55) Add the so-called Fear Of Missing Out (FOMO) to that and it <u>quickly</u> becomes clear that social media has the ability to wreck your night's rest.

STANCE MARKERS can also be used to signal an explanation through 'namely,' 'after all,' and 'then.'

(56) The conclusion <u>then</u>, is: ...
(57) It is <u>after all</u> not bad to miss something every now and then.

Sometimes a writer does not want to fully commit to a statement and will use 'almost,' 'a sort of,' or 'seem like.'

(58) If you look at it that way, social media doesn't <u>seem</u> that social.
(59) There is a <u>sort of</u> urge to constantly be available and online.

A reason is signaled by using 'so.'

(60) <u>So</u>, it comes as no surprise that social media influences the sleep of these teens.

Giving focus by using 'especially' signals that something is mostly true for a specific situation.

(61) Therefore, the sleep quality and the number of hours of sleep deteriorate, <u>especially</u> with night-time use of social media.

7.8   ANALYSIS EXAMPLE 5: REFERENCES TO THE READER

The results of the analysis for the strategy REFERENCES TO THE READER, part of the theme Engagement, are shown in Table 7.5. Thirty-one percent of students include REFERENCES TO THE READER in their newspaper article. REFERENCES TO THE READER are used for the same type of goal as the strategy INCLUSIVE PRONOUNS: giving information or spurring the reader into taking action. Except in these instances, the references are used to present readers as part of the interaction, not to point to the writer and the reader together, as would be the case with INCLUSIVE PRONOUNS. REFERENCES TO THE READER consist of seven subtypes, six of which overlap with INCLUSIVE PRONOUNS: taking action, something happens to you/your body, daily action, direct address of the reader, the writer says that you do something, the writer says that you might do something, and you as a parent.

The most-used REFERENCE TO THE READER is a reference to spur the reader into taking action. In 38 of these 43 instances, the text is also coded as the strategy APPLIED IMPLICATIONS. Where INCLUSIVE PRONOUNS are used to signal collective action, REFERENCES TO THE READER are used to connect the action that should be taken by the reader and to make the action more concrete:

(62) So, next time <u>you</u> pick up <u>your</u> phone, or watch another Facebook video, think twice, and put it down.

Claims about what happens to readers' bodies are also made using REFERENCES TO THE READER. These claims usually show a strong connection to

Table 7.5   Results for REFERENCES TO THE READER

	Number of texts	Number of references
REFERENCES TO THE READER	43	200
Taking action	20	43
Something happens to you/your body	10	43
Daily action	10	41
Direct address of the reader	10	23
The writer says you do something	7	17
The writer says you might do something	7	10
You as a parent	4	11
Not further specified	6	12

results from the paper that are discussed in the text. By mentioning these results and using REFERENCES TO THE READER, the results are made of personal interest to the reader:

(63) Current research tells us that the time of using social media and emotional investment are also two important factors. When you spend time on your phone at night, you absorb a lot of radiation, and your sleepiness hormone is not produced or produced too little. Because of this, your sleep rhythm gets out of sync, and you have more difficulty falling asleep.

Like the use of INCLUSIVE PRONOUNS, the use of REFERENCES TO THE READER to describe daily actions is connected to phones or looking at your phone late at night. All instances are also coded as the strategy EXAMPLES FROM DAILY LIFE. Here, REFERENCES TO THE READER are used to describe how the daily action or behavior is performed by the reader:

(64) At night before going to bed, you must check your phone. That phone then doubles as an alarm clock, so you conveniently place it next to your head when you go to sleep. And the notifications that appear just before you fall asleep? Ah well, let's have a look.

Although all instances in the strategy contain some form of a REFERENCE TO THE READER, only a direct address of the reader explicitly addresses the reader of the text and no one else. In 11 instances, in Dutch 'u' or 'uw' is used, which are politeness forms for second-person pronouns. The use of 'u' and 'uw' enhances the idea of the reader being spoken to directly. Also note that Dutch has two forms for you: 'je' and 'jij.' 'Je' is unstressed and can also be used to refer to people in general. In the use of 'jij,' the referent is stressed and used to mean one specific person; in this context that is the reader of the text. This has implications for the coding of this corpus, as references using 'jij' can be seen as a direct address of the reader, whereas references using 'je' are not (and both had to be translated to 'you' in the examples, inconveniently). Most instances of a direct address are combined with a question to draw the attention of the reader; these questions often ask if a situation that is described in the text is also applicable for or familiar to the reader.

(65) Did <u>you</u> miss a Facebook or an Instagram post and are <u>you</u> suddenly not in the loop anymore? Do <u>you</u> have trouble sleeping? Are <u>you</u> experiencing a lot of stress?

The writer can also tell readers that the readers are doing something. Here, the writer describes a certain action that is connected to the reader by referencing them:

(66) <u>You</u> used to compare <u>yourself</u> with the people <u>you</u> knew and maybe two idols. Now <u>you</u> constantly see the best of the best because those people are of course extra popular on social media.

In other cases, the writer does not assume the reader to do something, instead adding a STANCE MARKER of doubt such as 'maybe' or 'might.' When doubt is used in combination with a REFERENCE TO THE READER, it introduces the possibility of the one, specific reader not performing the action. As a consequence of using a marker signaling doubt, claims can be a bit bolder without offending the reader.

(67) <u>You</u> most probably recognize FOMO in <u>yourself</u> or in someone around <u>you</u>.

Student writers use REFERENCES TO THE READER specifically to address the reader in their role as parents. As the academic text is about sleep deprivation and depression in teens, some writers have specifically addressed their text to parents of those teens. When using INCLUSIVE PRONOUNS, the writer and parents form a shared group, but when REFERENCES TO THE READER are used, they function to specifically make it clear to the reader that they are addressed in their role as a parent:

(68) <u>You</u> might recognize it as a parent, <u>your</u> child says he's going to bed at 23:00, but when <u>you</u> quickly go to the bathroom at 1:00, <u>you</u> see the lights in his room are still on.

7.9 Interpretation

For a more comprehensive overview of the insights that were garnered about the student writing in this corpus of first-year undergraduate texts, see Sterk et al. (2022), in which the use of all 34 strategies is briefly discussed and lessons for educational practice are drawn from the overarching insights that were produced in this analysis. In this chapter we have shown how the framework can be used in an educational setting to analyze student writing. For five strategies, we have shown *how often* first-year undergraduate students use the strategies in the framework in their newspaper articles, and *in what way* these strategies are used. As can be seen from the five strategies that are discussed in this chapter, there are specific ways in which student writers use each strategy. This means that there is enough similarity between the different texts in the corpus to categorize and generalize the way in which each strategy is being used. Each discussed strategy was used in multiple ways, or in other words, contains subcodes. This implies that there is no strict single way of using these strategies. For almost every strategy, outliers were found. This means that although these instances do adhere to the strategy as such, the specific framing of the strategy is not used multiple times. It should be noted that there is no 'right' or 'wrong' way to use strategies; this analysis was not normative in nature.

Through the analysis of five themes, interrelations between strategies are already becoming visible. This can be seen, for example, in the use of LEXICAL MENTION OF THE ORIGINAL RESEARCH in DESCRIBING THE METHOD or the MENTION OF STATISTICS as part of NOVELTY, CONTEXTUALIZATION, and PRESENTING RESULTS/CONCLUSIONS. These interrelations become even clearer when looking at the use of all 34 strategies. This implies that although strategies can be analyzed individually, they cannot be pulled apart completely. This is a consequence of the occurrence of overlapping strategies and the allowance within the framework of overlapping coding.

A possible future step is to compare the analytic outcomes from this specific corpus to those of other corpora. For example, this corpus could be compared to a corpus constructed from texts from students from another research field, to study the influence of disciplinary perspectives on science writing. Or the current corpus could be compared to a corpus of texts about the same academic texts, and written by the same students, but later in time, for example at the end of their undergraduate studies. This could give insight into how the use of strategies in science writing develops over time. It could also be compared to a corpus of texts written by professional science journalists, to study the differences between those who are professionally trained to write and those who are not.

What You Have Learned in This Chapter

- This chapter presents the analysis of a corpus of 140 newspaper articles based on a single academic source text written by first-year undergraduate liberal education students.
- For five strategies—DESCRIBING THE METHOD, APPLIED IMPLICATION, MENTION OF STATISTICS, STANCE MARKERS, and REFERENCES TO THE READER—quantitative data were generated about the amount of use, and in a second analytic step, the way each strategy was used was described.
- DESCRIBING THE METHOD is achieved through mentioning the participants, research question/goal/hypothesis, measured constructs, and materials. Mentions of the original research are also included.
- There are two types of APPLIED IMPLICATIONS. Specific implications discuss further-reaching consequences of research findings and are presented on the level of teens, mental health, society, and parents. A call to action is used to nudge the reader into taking action and to discuss conscious choices in social media or using a mobile phone less.
- MENTIONS OF STATISTICS show issues teens have and behavior of teens. STATISTICS are re-presented from the results section as well as the introduction of the academic source text but are almost exclusively used in the student writing to denote results.
- STANCE MARKERS are mainly used to comment on the research, its findings, or its implications.
- REFERENCES TO THE READER are used to give information or to spur the reader into taking action. These references are employed to include the reader as part of the interaction.

REFERENCES

Sterk, F. M., Van Goch, M. M., Burke, M., & Van der Tuin, I. (2022). Baseline assessment in writing research: A case study of popularization discourse in first-year undergraduate students. *Journal of Writing Research, 14*(1), 35–76. https://doi.org/10.17239/jowr-2022.14.01.02

Woods, H. C., & Scott, H. (2016). #Sleepyteens: Social media use in adolescence is associated with poor sleep quality, anxiety, depression and low self-esteem. *Journal of Adolescence, 51,* 41–49. https://doi.org/10.1016/j.adolescence.2016.05.008

CHAPTER 8

Framing Frameworks: Final Considerations about Framework Development

Abstract In this concluding chapter, we look back on the contents of *Re-presenting Research* as a whole. We offer insight into the theoretical and applied insights that are generated through this book, which concern the framework itself, compliance with the aims that were set for the framework, the genre of popularization discourse as a whole, and applied insights. Options for further research are discussed, which include branching into other modes of popularization, focusing on the producer of popularization texts, the application of the framework in other settings, adding an evaluative component to the framework, and the construction of controversy in popularization discourse. Lastly, we offer some considerations for those readers wanting to develop their own analytical framework, whether it be to analyze popularization discourse or any other type of genre or text.

Keywords Insights • Implications • Further research • Text evaluation • Framework development

8.1 INTRODUCTION

Re-presenting Research has focused on the analysis of popularization discourse. In Chaps. 2 and 3, the theoretical and methodological backgrounds and underpinnings for popularization discourse and its analysis were put forward. In Chap. 4, we presented our analytical framework for

© The Author(s) 2023
F. M. Sterk, M. M. van Goch, *Re-presenting Research*,
https://doi.org/10.1007/978-3-031-28174-7_8

the analysis of popularization strategies in popularization discourse. The framework is constructed through the iterative analysis of a corpus of student-written newspaper articles through which an a priori list of codes (or strategies) was improved. This code list was then validated using a corpus of professional science journalism texts. The analytical framework consists of 34 strategies in five themes, which are specified in Chap. 4 and identified through application remarks and further reading suggestions. In Chaps. 5 and 6 we showed the application of the framework on the level of the individual text and on a corpus level. In Chap. 7, we added an extra, inductive step, to show how the analytical framework can also be a starting point for the further categorizing of *how* a strategy is employed in a specific corpus of texts.

8.2   Theoretical and Applied Insights Generated Through This Book

Previous studies into popularization strategies have offered insight into the textual features used in a specific subgenre of popularization discourse. Yet none of these studies has turned these insights into a framework that can be used in subsequent studies or projects, in other words, a framework that is applicable to code other texts or corpora than the one(s) under discussion in that specific academic paper. Our framework is the first *analytical* framework for popularization discourse. As such, it adds to the methodology of text analysis of popularization discourse and presents an addition to the methodological options available in the fields of science communication, discourse analysis, and communication studies.

Our framework is compliant with the four requirements of a proper analytical framework that were discussed in Chap. 4. The framework is usable in any subgenre of popularization discourse and any disciplinary, multidisciplinary, or interdisciplinary setting because the presented strategies are general enough to overarch constraints imposed by subgenres of popularization or disciplinary boundaries. Furthermore, as the framework is developed by using two raters and validated through a high inter-rater reliability, and since it includes application remarks for each strategy, it is workable and reliable in analyses using multiple raters. The framework also contains an explanation of each strategy to make it easy to apply and usable in other studies, or indeed by professionals in the fields of science communication and science journalism, or teachers and students in educational settings. As such, our framework enables both quantitative and instrumental as well as qualitative and inductive analysis (see Kuckartz, 2019; Roberts, 2000): it can be used to show numerical values about *how often* each strategy is used, as well as categorical data about *in what way* each strategy is used.

Concurrent with the theoretical insights that were presented about popularization discourse as a whole, this book has also offered insights for the applied use of a framework for popularization discourse. In Chap. 6 we focused on professional writing and in Chap. 7 on student writing, which shows that the analytical framework can be applied to different corpora within the genre of popularization discourse. Even though a direct comparison between the two corpora is outside of the scope of the research that we have conducted, anecdotally, we were able to share the insight that the use of strategies was denser in professional writing compared to student writing.

8.3 Options for Further Investigation and Research

Where do we go next? This book has hopefully given readers insight into textual strategies in popularization discourse, in the set-up of (analytical) frameworks, and in popularization discourse as a whole—but of course, research is never finished. In our research, the focus has been firmly on textual representations of science communication and science journalism. As such, the analytical framework is only applicable to written text. It does not cover spoken, visual, and interactive science communication or science journalism—but the framework could very well be adapted to evaluate these types of communication. The analysis of these types of popularization therefore remains for future studies.

Furthermore, this book has only dealt with the *product* of popularization, in other words text. Other relevant and adjacent avenues of investigation could consider the *producer* and the *recipient* of popularization: the writer and the reader. Such studies could, for example, ask: Which competences—knowledge, skills, and attitudes—are needed for popularized writing? What should popularization writing training focus on, and which of the framework's strategies are trainable? To give an example: How can you learn to use humor in writing? Which aspects of metacognitive awareness do professional science writers employ while writing, compared to student writers? How do readers perceive science journalism texts: Do they recognize the framework's strategies, and do they rate the quality of texts that use many strategies higher than those that use fewer? And what does 'quality' mean in this regard? Do readers get a better understanding, or appreciation, of academic discourse through reading science journalism? Similarly, we have only focused on text analysis using strategies. Other relevant questions would be: What does a *good* science journalism text look like, or a *creative* text, or *trustworthy* text? Another interesting research direction would be to study text production, in other words to investigate the actual writing processes taking place to construct popularization texts.

Another avenue of research is to consider the application of the framework in other settings. What would happen if professional science journalists used the framework in their writing? This would mean that research would focus on the use of the framework on the production side of text, and not solely on the analysis side of text production. It would also entail a transdisciplinary component to the research, where academics and practitioners could work together in research about text production. On the production side of texts, there are also opportunities to explore the inherent tension that exists for science communicators and science journalists who are responsible for the translation of academic insights to a broad target audience: there is a fine line between presenting new insights in an understandable manner and oversimplification, and between focusing on newsworthiness and becoming inaccurate. In this line of research, the interrelation between science journalists and press officers should not be glossed over. Press officers working, for instance, at universities or research institutes often produce the first recontextualization and reformulation steps when writing a press release about new research. Science journalists often use these press releases to determine which stories to write about and as a stepping stone for their own text. Research about science journalism text production should therefore take both science journalists and press officers into consideration as producers of popularization discourse.

Furthermore, teachers could use the framework in an educational setting. The current framework is usable to teach about the production of popularization discourse. However, it lacks an evaluative component; the framework cannot give any insight into if a strategy is used 'properly' or 'effectively,' nor can it be used as a rubric to grade popularization writing. Future research could focus on the addition of an evaluative component and the adaptation of the framework into the form of a rubric.

Popularization discourse is often produced about controversial topics or can give rise to societal debate about research topics that lead to controversy. Although controversy as such has been outside of the scope of this book, it should be noted that it is an integral part of the genre of popularization. An unexpected finding in our research is that controversy is hardly ever found on the level of a single text, which is why there is no strategy in the framework dedicated to it. Perhaps controversy is constructed intertextually, or in the interaction between discourse and society, or between author/text and reader, but more research is needed to provide better insight into this matter.

8.4 Considerations for Readers Who Want to Develop Their Own Framework

In *Re-presenting Research*, we showed how we constructed our analytical framework and how it can be applied in practice. For those readers wanting to develop their own framework, we would like to share some practical considerations that we developed throughout our research.

1. Do not try to develop a framework on your own. One of the most valuable aspects in the construction of our framework, we found, was the opportunity to thoroughly discuss the text analysis that we were working on, particular codes or texts with which we struggled, and adaptations to the framework. Discussion also offers you a clear reflection of your own frame of reference; the way you code a text fragment might be totally clear to you but could come under scrutiny by your fellow coders. Having discussions about how you code and why you do so is therefore very insightful and necessary in the development of a framework that has the aim of being as objective as possible.

2. Start working from an existing framework, not from scratch. Starting from an existing framework enables you to have a base to work from—if themes or strategies/codes in this existing framework do not match up with the data you are working with, this will become clear soon enough, and it can offer a stepping stone for discussions about adaptations of the framework.

3. Use a big corpus of texts to construct and validate your framework, but do not go overboard. As was shown by August et al. (2020),

coding of very big samples of text (that is, a 128,000-document corpus) is only really possible by employing data science techniques. Even then, hand-coding of a portion of those texts is needed in the validation of the computational analysis. The biggest hand-coded corpus that we encountered while researching popularization discourse frameworks was 337 texts, once again from the research by August et al. (2020)—though this analysis was performed on the sentence level. The chosen level of detail in a framework, and subsequently, the analysis of texts, also influences the number of texts that is realistically codable with the available number of analysts and within the set timeframe of the research. Keep in mind that there are two limiting factors in text analysis: it is necessary to achieve saturation (that is, you have coded all that needs to be coded according to your own coding parameters or theoretical framework) versus the time that is available. The higher the level of detail you include in your framework, the smaller the number of texts that are codable within a certain timeframe while still maintaining saturation of data.

4. Work in iterative rounds. In our research, we needed seven coding rounds to construct an analytical framework that we were happy with, that produced a very high level of agreement, and that was a true reflection of the texts that we were coding. It is also hugely beneficial for the development of the framework to code new texts from the same corpus in the different iterative rounds. Of course, no two texts are the same, and a new batch of texts often means that you are presented with new coding problems that call for a further adaptation of the framework—until at some point you reach code saturation. Also make sure to code the same texts as your fellow analysts, at least in the construction phase of the framework, to allow for true discussion about the coding.

5. Document everything. This includes difficulties in coding, differences in coding, topics that were under discussion among analysts, and changes made to the framework. The decisions you make might make sense to you now, but 'future you' will be thankful to know why a certain decision was made or how the framework was adapted.

6. Determine when your framework is considered to be a finished product. A framework is never truly finished; this is to say that further discussion about the framework and, potentially, adaptations to it is always possible. Even though our framework is presented as a 'finished product' in this book and is definitely highly applicable to

the genre that it was constructed for, further analysis of texts could always reveal more strategies that had previously not been encountered. To give a concrete example, as we noted in Chap. 5, the science journalist will also sometimes give a voice to researchers that were not part of the research project under discussion in the text. The voice of this other researcher is added either to give an explanation about difficult subject matter or to add an alternative opinion or interpretation. This textual feature is not captured well enough under the strategies GIVING THE RESEARCHER AN ACTIVE VOICE or GIVING THE NON-RESEARCHER AN ACTIVE VOICE. It *might* be coded as EXPLANATION, OPINION, or ADDITIONAL SOURCES. Alternatively, an extra strategy might need to be added to the framework. The same goes for the use of self-reference in a quote, which therefore cannot be coded as EXPLICIT SELF-REFERENCE. Here too, an additional strategy might be needed—but this is also dependent upon the question if and how adding an additional strategy would enrich our understanding of textual features in popularization discourse. All in all, the choice is yours to keep adapting a framework or to mark it as 'finished' when it can describe the chosen genre well enough.

8.5   Final Remarks

In 1986, Fahnestock was one of the first researchers to discuss popularization as a discourse. Over time, with the development of popularization discourse as a genre, implicit rules have formed about what can and cannot be done in popularization texts and, consequently, the genre has become rigid (as a sidebar, this process happens for all genres, not just popularization discourse). For popularization discourse, this has meant that the focus is firmly on news value and research results. This is what Fahnestock referred to as 'the wonder' and 'the application' as the two main themes in the discourse (Fahnestock, 1986). Generally, the produced insights are re-presented as 'facts' or 'the truth'—even though some of the main aims of popularization discourse are to show the audience the value of and need for scientific research, and to convince them that academic research is a reliable process. It would make sense, therefore, to give more attention to the way research is being conducted and scientific findings are produced, but methodology is hardly ever extensively or comprehensively discussed in these texts. The exception to the rule, here, is research where the methodology is very innovative or flashy. And this is where we once again return to the immune therapy research from the Rosalind Franklin Institute and the University of Reading that put Fifi the Llama—and, through Fifi, the methodology that was used—right in the center of their communication. But this is an exception to the current rules about what popularization discourse should look like. In the future, let us focus on alternative and creative ways to talk about research that do not solely focus on results.

What You Have Learned in This Chapter

- This book has shone a light on popularization discourse as a genre, as well as presenting the first analytical framework with which to analyze this type of discourse.
- Further investigation could focus on applications of the frameworks for other modes of popularization, on the producer or recipient of popularization discourse, on the application of the framework in the production of science journalism or in educational settings, on the addition of an evaluative component, or on the way that controversy surrounding research findings and popularization discourse is constructed.
- Considerations for developing a framework for text analysis are: do not try to develop the framework on your own, start working from an existing framework, use a big corpus of texts, work in iterative rounds, document everything, and determine when your framework is considered to be a finished product.

REFERENCES

August, T., Kim, L., Reinecke, K., & Smith, N. A. (2020). Writing strategies for science communication: Data and computational analysis. *Proceedings of the 2020 Conference on Empirical Methods in Natural Language Processing*, 5327–5344. https://doi.org/10.18653/v1/2020.emnlp-main.429

Fahnestock, J. (1986). Accommodating science: The rhetorical life of scientific facts. *Written Communication, 3*(3), 275–296. https://doi.org/10.1177/2F0741088386003003001

Kuckartz, U. (2019). Qualitative text analysis: A systematic approach. In G. Kaiser & N. Presmeg (Eds.), *Compendium for early career researchers in mathematics education (pp. 181-198)*. SpringerOpen. https://doi.org/10.1007/978-3-030-15636-7_8

Roberts, C. W. (2000). A conceptual framework for quantitative text analysis: On joining probabilities and substantive inferences about texts. *Quality & Quantity, 34*(3), 259–274. https://doi.org/10.1023/A:1004780007748

GLOSSARY

Term	Explanation
Academic discourse	Discourse that pertains to academic endeavors, for example academic articles or research reports.
Codes or categories	A code or category is part of a category system, that is, markers used to categorize text. They can be factual, thematic, evaluative, analytical, theoretical, natural, or formal.
Coding	A process that attaches a code to text.
Corpus	A corpus is a collection of texts surrounding the same genre, text type, or theme.
Framework	An analytic codebook that displays textual features (see: strategy) and their properties.
Popularization	The process of constructing popularization discourse.
Popularization discourse	Discourse in which academic insights are communicated to a broad, non-expert audience in understandable language and in the context of everyday life.
Research communication	The communication of academic insights from the humanities and social sciences (SSH field) to a broad, non-expert audience. For communication from the STEM field, see science communication.
Rubric	An analytic table that shows assessment criteria and grading points and can be used to assess student performance.
Science communication	The communication of academic insights from the natural sciences/ STEM field to a broad, non-expert audience. For communication from non-STEM fields, see research communication.

(continued)

F. M. Sterk, M. M. van Goch, *Re-presenting Research*,
https://doi.org/10.1007/978-3-031-28174-7

(continued)

Term	Explanation
Science journalism	The communication of academic insights from any field (though usually natural sciences/STEM) to a broad, non-expert audience, performed by a journalist.
Strategy	A theory-driven, communicative goal that consists of lexical to multi-sentence features.
Text analysis	A research method in which insights about written text are produced by turning text into data through the process of coding.
Qualitative text analysis	A type of text analysis that reduces the complexity of text into categories/codes to develop a category system/coding frame.
Quantitative text analysis	A type of text analysis that scores the occurrence of categories, sometimes to then perform statistical analysis.

Index[1]

---

[1] Note: Page numbers followed by 'n' refer to notes.

© The Author(s) 2023
F. M. Sterk, M. M. van Goch, *Re-presenting Research*,
https://doi.org/10.1007/978-3-031-28174-7

137